U.S. Department of Transportation
National Highway Traffic Safety Administration

DOT HS 810 767　　　　　　　　　　　　　　　　　　　　　　　　April 2007

Pre-Crash Scenario Typology for Crash Avoidance Research

This document is available to the public from the National Technical Information Service, Springfield, Virginia 22161

This publication is distributed by the U.S. Department of Transportation, National Highway Traffic Safety Administration, in the interest of information exchange. The opinions, findings and conclusions expressed in this publication are those of the author(s) and not necessarily those of the Department of Transportation or the National Highway Traffic Safety Administration. The United States Government assumes no liability for its content or use thereof. If trade or manufacturer's names or products are mentioned, it is because they are considered essential to the object of the publication and should not be construed as an endorsement. The United States Government does not endorse products or manufacturers.

REPORT DOCUMENTATION PAGE

Form Approved
OMB No. 0704-0188

Public reporting burden for this collection of information is estimated to average 1 hour per response, including the time for reviewing instructions, searching existing data sources, gathering and maintaining the data needed, and completing and reviewing the collection of information. Send comments regarding this burden estimate or any other aspect of this collection of information, including suggestions for reducing this burden, to Washington Headquarters Services, Directorate for Information Operations and Reports, 1215 Jefferson Davis Highway, Suite 1204, Arlington, VA 22202-4302, and to the Office of Management and Budget, Paperwork Reduction Project (0704-0188), Washington, DC 20503.

1. AGENCY USE ONLY (Leave blank)	2. REPORT DATE April 2007	3. REPORT TYPE AND DATES COVERED Final Report, 2004 - 2006 October 2003 – October 2005
4. TITLE AND SUBTITLE Pre-Crash Scenario Typology for Crash Avoidance Research		5. FUNDING NUMBERS PPA # HS-19
6. AUTHOR(S) Wassim G. Najm, John D. Smith, and Mikio Yanagisawa		
7. PERFORMING ORGANIZATION NAME(S) AND ADDRESS(ES) U.S. Department of Transportation Research and Innovative Technology Administration John A. Volpe National Transportation Systems Center Cambridge, MA 02142		8. PERFORMING ORGANIZATION REPORT NUMBER DOT-VNTSC-NHTSA-06-02
9. SPONSORING/MONITORING AGENCY NAME(S) AND ADDRESS(ES) U.S. Department of Transportation National Highway Traffic Safety Administration 400 7th St. SW Washington, DC 20590		10. SPONSORING/MONITORING AGENCY REPORT NUMBER DOT HS 810 767

11. SUPPLEMENTARY NOTES

12a. DISTRIBUTION/AVAILABILITY STATEMENT This document is available to the public through the National Technical Information Service, Springfield, Virginia 22161.	12b. DISTRIBUTION CODE

13. ABSTRACT (Maximum 200 words)

This report defines a new pre-crash scenario typology for crash avoidance research based on the 2004 General Estimates System (GES) crash database, which consists of pre-crash scenarios depicting vehicle movements and dynamics as well as the critical event immediately prior to a crash. This typology establishes a common vehicle safety research foundation for public and private organizations, which will allow researchers to determine which traffic safety issues should be of first priority to investigate and to develop concomitant crash avoidance systems. Its main objectives are to identify all common pre-crash scenarios of all police-reported crashes involving at least one light vehicle (i.e., passenger car, sports utility vehicle, van, minivan, and light pickup truck); quantify their severity in terms of frequency of occurrence, economic cost, and functional years lost; portray each scenario by crash contributing factors and circumstances in terms of the driving environment, driver, and vehicle; and provide nationally representative crash statistics that can be annually updated using national crash databases such as GES. This new typology includes 37 pre-crash scenarios accounting for approximately 5,942,000 police-reported light-vehicle crashes, an estimated economic cost of 120 billion dollars, and 2,767,000 functional years lost. These statistics do not incorporate data from non-police-reported crashes.

14. SUBJECT TERMS General Estimates System, pre-crash scenarios, vehicle safety research, crash avoidance research, crash frequency, economic cost, functional years lost			15. NUMBER OF PAGES 128
			16. PRICE CODE
17. SECURITY CLASSIFICATION OF REPORT Unclassified	18. SECURITY CLASSIFICATION OF THIS PAGE Unclassified	19. SECURITY CLASSIFICATION OF ABSTRACT Unclassified	20. LIMITATION OF ABSTRACT

NSN 7540-01-280-5500

Standard Form 298 (Rev. 2-89)
Prescribed by ANSI Std. 239-18
298-102

PREFACE

The National Highway Traffic Safety Administration (NHTSA), in conjunction with the Research and Innovative Technology Administration's Volpe National Transportation Systems Center (Volpe Center), conducts vehicle safety research in crash avoidance and crashworthiness. In particular, extensive analyses have been performed to define the crash and injury problems, identify intervention opportunities, assess the state-of-the-art technology for crash avoidance and injury mitigation systems, and estimate potential safety benefits of promising systems. This research supports NHTSA's mission to save lives, prevent injuries, and reduce health care and other economic costs associated with motor vehicle crashes.

This report presents results obtained from the analysis of the 2004 General Estimates System crash database. It describes a new typology of pre-crash scenarios leading to all police-reported crashes that involve at least one light vehicle (e.g., passenger car, sports utility vehicle, van, minivan, and light pickup truck).

Authors of this report are Wassim G. Najm, John D. Smith, and Mikio Yanagisawa of the Volpe Center.

The authors acknowledge the technical contribution and cooperation from Dr. David L. Smith and Mr. Ray Resendes of NHTSA. This acknowledgement is also extended to Mr. Richard Deering of General Motors for his technical support and desire for cooperation between the automotive industry and NHTSA in vehicle safety research. Also acknowledged are the technical staffs from the Crash Avoidance Metrics Partnership.

METRIC/ENGLISH CONVERSION FACTORS

ENGLISH TO METRIC	METRIC TO ENGLISH
LENGTH (APPROXIMATE)	**LENGTH** (APPROXIMATE)
1 inch (in) = 2.5 centimeters (cm)	1 millimeter (mm) = 0.04 inch (in)
1 foot (ft) = 30 centimeters (cm)	1 centimeter (cm) = 0.4 inch (in)
1 yard (yd) = 0.9 meter (m)	1 meter (m) = 3.3 feet (ft)
1 mile (mi) = 1.6 kilometers (km)	1 meter (m) = 1.1 yards (yd)
	1 kilometer (km) = 0.6 mile (mi)
AREA (APPROXIMATE)	**AREA** (APPROXIMATE)
1 square inch (sq in, in^2) = 6.5 square centimeters (cm^2)	1 square centimeter (cm^2) = 0.16 square inch (sq in, in^2)
1 square foot (sq ft, ft^2) = 0.09 square meter (m^2)	1 square meter (m^2) = 1.2 square yards (sq yd, yd^2)
1 square yard (sq yd, yd^2) = 0.8 square meter (m^2)	1 square kilometer (km^2) = 0.4 square mile (sq mi, mi^2)
1 square mile (sq mi, mi^2) = 2.6 square kilometers (km^2)	10,000 square meters (m^2) = 1 hectare (ha) = 2.5 acres
1 acre = 0.4 hectare (he) = 4,000 square meters (m^2)	
MASS - WEIGHT (APPROXIMATE)	**MASS - WEIGHT** (APPROXIMATE)
1 ounce (oz) = 28 grams (gm)	1 gram (gm) = 0.036 ounce (oz)
1 pound (lb) = 0.45 kilogram (kg)	1 kilogram (kg) = 2.2 pounds (lb)
1 short ton = 2,000 pounds (lb) = 0.9 tonne (t)	1 tonne (t) = 1,000 kilograms (kg)
	= 1.1 short tons
VOLUME (APPROXIMATE)	**VOLUME** (APPROXIMATE)
1 teaspoon (tsp) = 5 milliliters (ml)	1 milliliter (ml) = 0.03 fluid ounce (fl oz)
1 tablespoon (tbsp) = 15 milliliters (ml)	1 liter (l) = 2.1 pints (pt)
1 fluid ounce (fl oz) = 30 milliliters (ml)	1 liter (l) = 1.06 quarts (qt)
1 cup (c) = 0.24 liter (l)	1 liter (l) = 0.26 gallon (gal)
1 pint (pt) = 0.47 liter (l)	
1 quart (qt) = 0.96 liter (l)	
1 gallon (gal) = 3.8 liters (l)	
1 cubic foot (cu ft, ft^3) = 0.03 cubic meter (m^3)	1 cubic meter (m^3) = 36 cubic feet (cu ft, ft^3)
1 cubic yard (cu yd, yd^3) = 0.76 cubic meter (m^3)	1 cubic meter (m^3) = 1.3 cubic yards (cu yd, yd^3)
TEMPERATURE (EXACT)	**TEMPERATURE** (EXACT)
[(x-32)(5/9)] F = y C	[(9/5) y + 32] C = x F

QUICK INCH - CENTIMETER LENGTH CONVERSION

QUICK FAHRENHEIT - CELSIUS TEMPERATURE CONVERSION

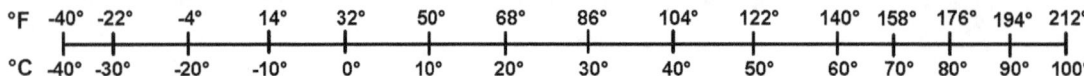

For more exact and or other conversion factors, see NIST Miscellaneous Publication 286, Units of Weights and Measures. Price $2.50 SD Catalog No. C13 10286

TABLE OF CONTENTS

EXECUTIVE SUMMARY .... *v*

1. INTRODUCTION *1*
    1.1. 44-Crashes Typology 2
    1.2. Pre-Crash Scenarios Typology 6
    1.3. Report Outline 7

2. IDENTIFICATION OF NEW PRE-CRASH SCENARIO TYPOLOGY *8*
    2.1. Scenario Coding Schemes 8
    2.2. Crash Contributing Factors and Circumstances 9
    2.3. Societal Harm Measures 11

3. DESCRIPTION OF LIGHT-VEHICLE CRASHES *14*
    3.1. Crash Severity 14
    3.2. Crash Breakdown by Number of Vehicles Involved Per Crash 14
    3.3. Contributing Factors and Circumstances of Light-Vehicle Crashes 15

4. DETAILS OF NEW PRE-CRASH SCENARIO TYPOLOGY *19*
    4.1. Single-Vehicle Pre-Crash Scenarios 19
    4.2. Two-Vehicle Pre-Crash Scenarios 21
    4.3. Multi-Vehicle (> 2) Pre-Crash Scenarios 22
    4.4. All Light-Vehicle Pre-Crash Scenarios 24
    4.5. Statistical Description of All Light-Vehicle Pre-Crash Scenarios 27

5. MAPPING TO NEW PRE-CRASH SCENARIO TYPOLOGY *65*
    5.1. Mapping of a Sample of Police-Reported Crashes 65
    5.2. Mapping of 44 Crashes 67
    5.3. Mapping of Crash Types 69

6. CONCLUSIONS *71*

7. REFERENCES *73*

APPENDIX A. IDENTIFICATION CODES OF PRE-CRASH SCENARIOS USING THE GENERAL ESTIMATES SYSTEM *74*

APPENDIX B. CRASH CHARACTERISTICS OF PRE-CRASH SCENARIOS *78*

LIST OF TABLES

Table 1. List of 44 Crash Scenarios. 3

Table 2. List of Pre-Crash Scenario Based on NASS Variables 6

Table 3. MAIS Levels and Unit Costs in 2000 Dollars 12

Table 4. Functional Years Lost by MAIS Per-Unit Basis...13

Table 5. Injury Severity Comparison Between Light-Vehicle and All-Vehicle Crashes 14

Table 6. Comparison of Crash Severity Between Light-Vehicle and All-Vehicle Crashesby Number of Vehicles Involved per Crash............................15

Table 7. Driving Environment Statistics of Light-Vehicle Crashes................................ 16

Table 8. Driver Factors Statistics of All Light-Vehicle Drivers 17

Table 9. Vehicle Factor Statistics of All Light Vehicles...18

Table 10. Pre-Crash Scenarios of Single-Vehicle Light-Vehicle Crashes........................22

Table 12. Pre-Crash Scenarios of Multi-Vehicle Light-Vehicle Crashes 23

Table 13. Pre-Crash Scenarios of All Light-Vehicle Crashes..25

Table 14. Ranking of Light-Vehicle Pre-Crash Scenarios by Economic Cost 26

Table 15. Ranking of Light-Vehicle Pre-Crash Scenarios by Functional Years Lost 27

Table 16. Mapping of a Sample of Crash Reports to New Pre-Crash Scenario Typology 66

Table 17. Mapping of 44 Crashes to New Pre-Crash Scenario Typology 68

Table 18. Mapping of Crash Types to New Pre-Crash Scenario Typology.....................70

LIST OF FIGURES

Figure 1. Distribution of Light-Vehicle and All-Vehicle Crashes by Number of Vehicles Involved per Crash 15

EXECUTIVE SUMMARY

This report defines and statistically describes a new pre-crash scenario typology for light vehicles (i.e., passenger car, sports utility vehicle, minivan, van, and light pickup truck) based on the 2004 General Estimates System (GES) crash database. This new typology consists of pre-crash scenarios that depict vehicle movements and dynamics as well as the critical event occurring immediately prior to a crash. The goal of this typology is to establish a common vehicle safety research foundation for public and private organizations, which will allow researchers to determine which traffic safety issues should be of first priority to investigate and to develop concomitant crash avoidance systems. Its main objectives are to identify all common pre-crash scenarios of all police-reported crashes involving at least one light vehicle; quantify their severity in terms of frequency of occurrence, economic cost, and functional years lost; portray each scenario by crash contributing factors and circumstances in terms of the driving environment, driver, and vehicle; and provide nationally representative crash statistics that can be annually updated using GES and the Crashworthiness Data System (CDS) crash databases.

The following 37 pre-crash scenarios, including "other", comprise the new typology:

1	Vehicle Failure
2	Control Loss With Prior Vehicle Action
3	Control Loss Without Prior Vehicle Action
4	Running Red Light
5	Running Stop Sign
6	Road Edge Departure With Prior Vehicle Maneuver
7	Road Edge Departure Without Prior Vehicle Maneuver
8	Road Edge Departure While Backing Up
9	Animal Crash With Prior Vehicle Maneuver
10	Animal Crash Without Prior Vehicle Maneuver
11	Pedestrian Crash With Prior Vehicle Maneuver
12	Pedestrian Crash Without Prior Vehicle Maneuver
13	Pedalcyclist Crash With Prior Vehicle Maneuver
14	Pedalcyclist Crash Without Prior Vehicle Maneuver
15	Backing Up Into Another Vehicle
16	Vehicle(s) Turning – Same Direction
17	Vehicle(s) Parking – Same Direction
18	Vehicle(s) Changing Lanes – Same Direction
19	Vehicle(s) Drifting – Same Direction
20	Vehicle(s) Making a Maneuver – Opposite Direction
21	Vehicle(s) Not Making a Maneuver – Opposite Direction
22	Following Vehicle Making a Maneuver

23	Lead Vehicle Accelerating
24	Lead Vehicle Moving at Lower Constant Speed
25	Lead Vehicle Decelerating
26	Lead Vehicle Stopped
27	Left Turn Across Path From Opposite Directions at Signalized Junctions
28	Vehicle Turning Right at Signalized Junctions
29	Left Turn Across Path From Opposite Directions at Non-Signalized Junctions
30	Straight Crossing Paths at Non-Signalized Junctions
31	Vehicle(s) Turning at Non-Signalized Junctions
32	Evasive Action With Prior Vehicle Maneuver
33	Evasive Action Without Prior Vehicle Maneuver
34	Non-Collision Incident
35	Object Crash With Prior Vehicle Maneuver
36	Object Crash Without Prior Vehicle Maneuver
37	Other

- Vehicle Action refers to a vehicle decelerating, accelerating, starting, passing, parking, turning, backing up, changing lanes, merging, and successful corrective action to a previous critical event.
- Vehicle Maneuver denotes passing, parking, turning, changing lanes, merging, and successful corrective action to a previous critical event.

Pre-crash scenarios listed above accounted for approximately 5,942,000 police-reported crashes involving at least one light vehicle, and resulted in an estimated economic cost of $120 billion and 2,767,000 functional years lost. These statistics do not incorporate data from non-police-reported crashes. Excluding "other" scenario, this new pre-crash scenario typology represents about 99.4 percent of all light-vehicle crashes.

Pre-crash scenarios of this new typology were ranked using three measures: crash frequency, functional years lost, and economic cost. Table ES-1 lists the dominant pre-crash scenarios emerging from the top five scenarios in each of the three measures. Ranking by crash frequency, the five most frequent scenarios accounted for 45 percent of all police-reported light-vehicle crashes. Ranking by functional years lost, the top five scenarios resulted in 49 percent of all years lost. Ranking by economic cost, the top five scenarios contributed to 46 percent of all cost associated with light-vehicle crashes. As seen in Table ES-1, the three most dominant scenarios are:

1. Control loss without prior vehicle action
2. Lead vehicle stopped
3. Road edge departure without prior vehicle maneuver

Table ES-1. Dominant Pre-Crash Scenarios

Scenario	Occurrence		Functional Years Lost		Direct Economic Cost	
	Rank	Frequency	Rank	Years	Rank	Cost ($)
Control Loss Without Prior Vehicle Action	2	529,000	1	478,000	1	15,796,000,000
Lead Vehicle Stopped	1	975,000	3	240,000	2	15,388,000,000
Road Edge Departure Without Prior Vehicle Maneuver	5	334,000	2	270,000	3	9,005,000,000
Vehicle(s) Turning at Non-Signalized Junctions	3	435,000			4	7,343,000,000
Straight Crossing Paths at Non-Signalized Junctions			5	174,000	5	7,290,000,000
Lead Vehicle Decelerating	4	428,000				
Vehicle(s) Not Making a Maneuver – Opposite Direction			4	206,000		

1. INTRODUCTION

A number of crash typologies have been developed over the years in support of vehicle safety research. Crash typologies provide an understanding of distinct crash types and scenarios and explain why they occur. They serve as a tool to identify intervention opportunities, set research priorities and direction in technology development, and evaluate the effectiveness of selected crash countermeasure systems. Recently, two crash typologies have been widely used for crash avoidance research in support of the Intelligent Vehicle Initiative (IVI) within the U.S. Department of Transportation's (USDOT) Intelligent Transportation Systems program: 44-crashes and pre-crash scenarios.

The 44-crashes typology has been developed by General Motors (GM) and adopted by automakers for the design, development, and benefits assessment of potential crash countermeasure technologies [1, 2]. This typology identified very specific crash scenarios representing all collisions in the United States and investigated the causes associated with each crash scenario using the 1991 General Estimates System (GES) crash database and samples of 1990-1991 police-reported crashes from Michigan and North Carolina. Shortcomings of this typology include the limited study of State crash data and the amount of effort required to replicate the results using recent crash data.

USDOT has devised the pre-crash scenarios typology based primarily on pre-crash variables in the National Automotive Sampling System (NASS) crash databases including GES and the Crashworthiness Data System (CDS) [3]. This typology has been utilized to identify intervention opportunities, develop performance guidelines and objective test procedures, and estimate the safety benefits for IVI crash countermeasure systems. Single-vehicle and two-vehicle crashes of common crash types were analyzed to produce the list of representative pre-crash scenarios. Multi-vehicle (> 2) crashes were not included in the analysis. Some low-frequency crash types were also excluded such as vehicle failure, non-collision incidents, and evasive action scenarios. As a result, the pre-crash scenario typology did not represent 100 percent of all police-reported crashes.

This report defines a new typology of pre-crash scenarios for crash avoidance research, which combines crash information from both typologies mentioned above. This new typology consists of pre-crash scenarios that depict vehicle movements and dynamics as well as the critical event occurring immediately prior to crashes involving at least one light vehicle (i.e., passenger car, sports utility vehicle, van, minivan, and light pickup truck). The goal of this typology is to establish a common vehicle safety research foundation for public and private organizations, which will allow researchers to determine which traffic safety issues should be of first priority to investigate and to develop concomitant crash avoidance systems. Its main objectives are to:

1. Identify all common pre-crash scenarios of all police-reported crashes involving at least one light vehicle.
2. Quantify the severity of each pre-crash scenario in terms of frequency of occurrence, direct economic cost, and functional years lost.

3. Portray each scenario by crash contributing factors and circumstances in terms of the driving environment, driver, and vehicle.
4. Provide nationally representative crash statistics that can be annually updated using GES and CDS crash databases.

This report describes a new typology that comprises all scenarios in the 44-crashes and pre-crash scenarios typologies using the 2004 GES crash database [4].

1.1. 44-Crashes Typology

Table 1 lists the 44 crashes developed by GM using multiple crash data sources [1]. This typology described the national crash problem based on an analysis of crash involved vehicles and factors that may increase the likelihood of occurrence. Three distributions of crashes were defined using the frequency of occurrence, losses due to direct costs, and losses due to years of functional life lost. There were originally 100 crash scenarios, each representing about one percent of the entire crash problem. Some scenarios have been combined because of similarities, thus bringing the list down to 44 crash scenarios.

The 44-crashes typology was developed to give in a simplistic sense an understanding of crashes and to prioritize crash countermeasure development. It also helps address some of the obstacles associated with trying to predict field effectiveness using raw statistics:

- Double counting: Consider two crash prevention measures that are each 10-percent effective. If they influence totally different crashes, then together they are probably about 20-percent effective. However, if their benefit applies to exactly the same crashes, then together they are only about 10-percent effective. To claim these redundant countermeasures are more than 10-percent effective is double-counting. The 44-crashes typology helps prevent double counting.
- Complexity of crash statistics: Crash statistics may be confusing and may take a long time to process, which cannot be used efficiently by technologists. This typology was proposed as a simple problem definition.
- Inconsistency: Crash avoidance has no standard metric, like emissions and fuel economy. Crash avoidance needed a standard problem definition that spanned organizations and time.

As seen in Table 1, the definition of the 44 scenarios incorporates vehicle dynamics, vehicle movements, critical events, crash causes, and crash contributing factors. Specifics of some scenario descriptions are not represented by GES variables and codes, such as pedal miss and other details of causal factors.

Table 1. List of 44 Crash Scenarios

No.	Title	Scenario Definition
1	Struck Human	A pedestrian crossing a multi-lane roadway was struck by vehicle. The driver was looking for other vehicles and traffic controls, but did not see the pedestrian. This crash occurs more frequently in urban areas.
3	Struck Animal	A male driving home after dark on a rural two-lane country road in November struck a deer crossing the road. The weather is typically clear and the road is usually dry. The driver could not avoid hitting the deer.
9	Drowsy	The driver fell asleep and drifted off the right side of the road and struck a telephone pole. Witnesses say that there was no attempt to brake or steer away from the pole. The crash occurred in a rural area at night.
10	Aggressive, Departure	The male driver was driving too fast, as well as cutting in and out of traffic, maneuvering the vehicle to the limits of control. The driver lost control of the vehicle and went into a skid. The driver left the roadway and struck the guardrail and then a tree.
11	Slick Road Departure	The driver lost control while driving on an icy, wet road. The driver tried to bring the vehicle back under control by braking and steering. The vehicle spun out and came to rest in the ditch.
12	Rough Road Departure	Due to the patched and eroded condition of the road surface, the driver lost control of the vehicle and left the roadway.
13	Avoidance, Departure	The driver was alert and driving along a surface street. Suddenly something appeared in the driver's path (e.g., child, bicyclist, or animal). The driver slammed on the brakes and swerved to avoid the immediate threat. The vehicle drove over a curb and into an object.
18	Impaired, Departure	The young (under 25) male driver, who was legally impaired, was driving too fast. He lost control of the vehicle, which left the roadway and overturned. The crash occurred in a rural area between midnight and 2 a.m. on a weekend.
19	Back Into Object	Vehicle A was backing out of a driveway and struck Vehicle B that was parked along the side of the road. Driver A did not see the other vehicle.
22	Ran Red "T-Bone"	Driver ran the red light. The driver saw the light turn yellow but decided to continue through the intersection. The majority of these crashes occur during daylight hours in urban areas.
28	Slick Road, Ran Stop	As vehicle approached an intersection, the driver noticed the stop sign, applied the brakes hard, but slid on the wet pavement into crossing traffic. (This group does not include the condition where there is no sign.)
30	Inattentive, Ran Stop	An inattentive driver in a vehicle, heading north, did not see a stop sign (two-way only) and struck an eastbound vehicle on the passenger's side.
33	View Obstruction	A vehicle, at a two-way stop sign, could not see adequately down the road due to the hill. This vehicle pulled out and was struck on the driver's side by a lateral-crossing vehicle. This crash is most likely to occur in daylight in rural areas.
35	Looked but Didn't See	Vehicle A was turning right at a two-way stop sign. The driver did not see Vehicle B approaching from lateral direction as Vehicle A turned into the lane. Upon turning, Vehicle A was struck by Vehicle B.
37	Sirens	A police car, with lights and siren on, slowed to cross through an intersection with a red light. Another vehicle was on the crossing road and did not see the approaching police car.
38	Left Turn Clip	Vehicle A, in an attempt to turn left, cut the corner too sharply and clipped Vehicle B waiting at the intersection. Vehicle A began the turn too early and misjudged the distance between cars.

Table 1. List of 44 Crash Scenarios (Cont. 1)

No.	Title	Scenario Definition
40	Wrong Driveway	Driver A observed Vehicle B approaching with the right turn signal on. A assumed that B was turning into the driveway that A was turning out of and proceeded in front of B. B was not turning until the intersection and struck A in the side.
44	Wave to Go	From a driveway, Vehicle A was waiting to make a left turn, but full view of all lanes was not possible due to other traffic. Driver B stopped—leaving a gap—and waved driver A through in front of him. However, Driver C was unaware of this arrangement and crashed into the driver's side of Vehicle A.
47	Turn into Passer	An impatient driver, A, was following behind a slower vehicle, B. Driver A passed vehicle B. Driver B turned left as A was passing and collided with A.
48	Back into Roadway	Driver A backed vehicle into roadway. Driver A did not see vehicle B heading west.
52	Tailgate	Vehicle B was following Vehicle A too closely. Vehicle A had to stop quickly; B could not stop in time and rear-ended A.
56	Distracted, Rear	The driver of Vehicle A was reaching down to retrieve an item from the floor of the vehicle and did not notice that Vehicle B was stopped ahead.
58	Avoidance, Rear	Vehicle A observed traffic slowing in the curb lane. A decided to change lanes and go around slowing traffic. A changed lanes to the inside lane only to find Vehicle B stopped directly in front. Driver A could not stop and struck B in the rear. (This also includes cases of three cars in the same lane. The middle vehicle pulled out of the lane at the last moment leaving the rear-most vehicle to collide with the foremost.)
61	Pedal Miss	Driver A was attempting to stop behind Vehicle B when Driver A's foot missed the brake pedal and Vehicle A struck Vehicle B from behind.
62	Inattentive, Rear	A northbound vehicle, A, was stopped waiting at a red traffic signal in an urban area on a major artery. Another vehicle, B, coming from some distance behind, didn't notice that A was stopped and could not stop in time. (This crash includes a lead vehicle just stopping or lead vehicle turning.)
64	Stutter Stop	A stopped vehicle, A, was looking left and right down a cross road waiting for traffic to clear before proceeding. Another driver, B, waiting behind A was also checking crossing traffic. Vehicle A started to go, decided that it wasn't safe, and abruptly stopped. Driver B, who had been watching traffic, thought that A had moved on and proceeded. Driver B rear-ended driver A.
66	Aggressive, Rear	Vehicle A was stopped in traffic. Driver B (at a distance from A) was driving too fast. By the time B realized he/she needed to stop, it was not possible.
68	Maintenance	Vehicle A was stopped prior to turning when struck by Vehicle B. Driver B stated that the brakes failed to stop the car. Vehicle B was an older vehicle (more than six years). (The failure is usually a maintenance problem.)
74	Slick Road, Rear	Vehicle A was braking for stopped traffic. Driver B, coming from some distance behind A, saw the brake lights. When B braked the road was very slick. B did not stop and struck A in the rear.
75	Passing Clip	Vehicle A, in an attempt to pass vehicle B, cut around B, but too closely. Driver A misjudged the distance between cars and clipped the corner of B.

Table 1. List of 44 Crash Scenarios (Cont. 2)

No.	Title	Scenario Definition
76	Lane Change Right	Driver of Vehicle A looked for traffic before changing lanes to the right on a four-lane road. The driver did not see Vehicle B in the curb lane. Vehicle B braked and steered to avoid Vehicle A.
78	Visibility, Rear	Driver A could not see well due to the blowing snow (whiteout conditions). Vehicle B was in front of A and traveling in the same direction. B had to brake for stopped traffic ahead and A did not notice the brake lights.
79	Lane Change Left	Driver of Vehicle A looked for traffic before changing lanes to the left on a four-lane road. The driver did not see Vehicle B in the next lane. Vehicle B had no time to react and nowhere to go to avoid Vehicle A.
80	Lane Change, Rear	Vehicle A saw Vehicle B approaching in the next lane. A determined that B was far enough back that A could change lanes. Driver A misjudged the distance and speed of Vehicle B. Driver B pressed the brake hard but was unable to stop and struck A from behind. Vehicle C could not stop and struck B from behind.
82	Back Track	Front Vehicle A stopped too far out in an intersection. Driver A did not see Vehicle B and backed up to allow other traffic through, striking vehicle B.
83	U-Turn	Vehicle A and vehicle B were both heading in the same direction on a multi-lane road in different lanes. B attempted to turn from the curb lane across the path of A onto a side street. Driver A struck illegally turning B in the driver's side.
91	Inexperience, Departure	Driver A was having a difficult time controlling the vehicle on the slippery road. The driver lost control of the vehicle while starting into a curve and applied the brakes. The vehicle crossed into the opposite direction traffic and collided head-on with Vehicle B. (This often involves a new driver or a driver who lacks experience on a roadway with a low coefficient of friction.)
92	Impaired, Head-on	A young male driver A, who was legally impaired, was driving too fast. He lost control of the vehicle, crossed the centerline, and struck an approaching vehicle head-on. The crash occurred in a rural area between midnight and 2 a.m. on a weekend.
93	Slick Road, head-on	Vehicle A attempted to stop at an intersection, but because of the slick road, lost control of the vehicle. Vehicle B was approaching head-on in the opposite direction and was struck by A.
94	Run Red Into Left Turner	A northbound vehicle, A, was waiting to make a left turn. The light changed and the northbound vehicle began to turn left. A southbound driver, B, accelerated hard, hoping to make the light and struck Vehicle A.
96	Misjudgment, Left Turn	Vehicle A was waiting to turn left. Driver A observed B approaching from the opposite direction, but thought there was enough time to complete the left turn. Driver A misjudged vehicle B's distance and was struck by Vehicle B.
99	View Obstructed Left Turn	Vehicle A was stopped in the left lane of a four-lane road, facing north, waiting to complete a left turn. Vehicle C was also stopped in the left lane in the opposite direction waiting to complete a left turn. Driver A, able to see past C only a short distance, thought it was clear and completed the turn. Vehicle B, in the curb lane adjacent to Vehicle C was traveling south at the posted speed limit and struck Vehicle A head-on.
100	Miscellaneous	This is a miscellaneous assortment that could not be classified as any of the other previously mentioned crash descriptions.
101	New	This is a crash that may not have occurred without the introduction of a new safety technology. The driver may have used the new technology for increased mobility rather than an increase in safety as intended. A crash may evolve to another type under the driver's control rather than becoming eliminated.

1.2. Pre-Crash Scenarios Typology

Table 2 lists the pre-crash scenarios developed by USDOT using primarily the Accident Type variable and the first two pre-crash variables in the NASS crash databases. These two pre-crash variables are the Movement Prior to Critical Event and Critical Event. The Accident Type variable categorizes the pre-crash situation. The Movement Prior to Critical Event variable records the attribute that best describes vehicle activity prior to the driver's realization of an impending critical event or just prior to impact if the driver took no action or had no time to attempt any evasive maneuver. The Critical Event variable identifies the circumstances that made the crash imminent.

The scenarios listed in Table 2 were identified within each of the following crash types: rear-end, off-road, lane change, crossing paths, opposite direction, backing, pedestrian, pedalcyclist, animal, and object crashes. Moreover, the identification of these scenarios was based on the analysis of single- and two-vehicle crashes. Crashes that involved more than two vehicles were excluded from the analysis due to the uncertainty and crosscutting among the various crash types as a result of associating the Accident Type variable with the pre-crash variables.

Table 2. List of Pre-Crash Scenarios Based on NASS Variables

No.	Scenario Definition
1	Animal: other
2	Animal: vehicle going straight and animal in road
3	Animal: vehicle negotiating a curve and animal in road
4	Off-road: single vehicle performing avoidance maneuver
5	Off-road: single vehicle going straight and departing road edge
6	Off-road: single vehicle going straight and losing control
7	Off-road: single vehicle initiating a maneuver and departing road edge
8	Off-road: single vehicle initiating a maneuver and losing control
9	Off-road: single vehicle negotiating a curve and departing road edge
10	Off-road: single vehicle negotiating a curve and losing control
11	Off-road: single vehicle and other loss of control
12	Off-road: single vehicle due to vehicle failure
13	Off-road: single vehicle and other road edge departure
14	Off-road: single vehicle with other/unknown
15	Off-road: backing
16	Off-road: no impact
17	Pedalcyclist: other/unknown
18	Pedalcyclist: vehicle going straight on crossing paths
19	Pedalcyclist: vehicle going straight on parallel paths
20	Pedalcyclist: vehicle starting in traffic lane on crossing paths
21	Pedalcyclist: vehicle turning left on crossing paths
22	Pedalcyclist: vehicle turning left on parallel paths
23	Pedalcyclist: vehicle turning right on crossing paths
24	Pedalcyclist: vehicle turning right on parallel paths
25	Pedestrian: other
26	Pedestrian: vehicle backing
27	Pedestrian: vehicle going straight and pedestrian crossing road
28	Pedestrian: vehicle going straight and pedestrian darting onto road

Table 2. List of Pre-Crash Scenarios Based on NASS Variables (Cont.)

No.	Scenario Definition
29	Pedestrian: vehicle going straight and pedestrian playing/working on Road
30	Pedestrian: vehicle going straight and pedestrian walking along road
31	Pedestrian: vehicle turning left and pedestrian crossing road
32	Pedestrian: vehicle turning right and pedestrian crossing road
33	Backing: at driveways
34	Backing: at intersections
35	Backing: other
36	Lane change: 2 vehicles going straight and 1 vehicle encroaching in same lane
37	Lane change: 2 vehicles going straight and 1 vehicle encroaching into another lane
38	Lane change: 1 vehicle going straight and another changing lanes
39	Lane change: 1 vehicle going straight and another entering or leaving parking position
40	Lane change: 1 vehicle going straight and another passing
41	Lane change: 1 vehicle going straight and another turning
42	Lane change: 2 vehicles in other combinations
43	Lane change: 1 vehicle passing and another turning
44	Opposite direction: control loss
45	Opposite direction: 2 vehicles going straight and 1 vehicle encroaching
46	Opposite direction: 2 vehicles going straight both in same lane
47	Opposite direction: 2 vehicles negotiating a curve and 1 vehicle encroaching
48	Opposite direction: 2 vehicles negotiating a curve both in same lane
49	Opposite direction: other/unknown
50	Opposite direction: involves 1 vehicle passing
51	Opposite direction: involves vehicle failure
52	Rear-end: following vehicle changing lanes
53	Rear-end: lead vehicle accelerating
54	Rear-end: lead vehicle changing lanes
55	Rear-end: lead vehicle decelerating
56	Rear-end: lead vehicle moving at constant, slower speed
57	Rear-end: lead vehicle stopped
58	Rear-end: other/unknown
59	Crossing paths: left turn across path from lateral direction (LTAP/LD)
60	Crossing paths: left turn across path from opposite direction (LTAP/OD)
61	Crossing paths: left turn into path (LTIP)
62	Crossing paths: other/unknown
63	Crossing paths: right turn across path from lateral direction (RTAP/LD)
64	Crossing paths: right turn into path (RTIP)
65	Crossing paths: straight crossing paths (SCP)

1.3. Report Outline

Following the introduction, this report delineates the approach used to identify and statistically describe the scenarios of the new pre-crash typology, and to estimate the societal cost measures of direct economic cost and functional years lost. This is followed by crash statistics of light-vehicle crashes. Afterwards, the new pre-crash typology is introduced and each of its scenarios is defined. After that, this report maps a sample of crash police reports, 44 crashes, and crash types to the new pre-crash typology. Finally, this report concludes with some comments about the overall analysis.

2. IDENTIFICATION OF NEW PRE-CRASH SCENARIO TYPOLOGY

GES was selected as the best available source for the identification and description of the new pre-crash scenario typology because it:

- Is nationally representative
- Is annually updated
- Contains the Accident Type variable and pre-crash variables that enable the identification of dynamically-distinct vehicle scenarios
- Features the availability of different sets of variables that describe the environmental and driving conditions at the time of the crash, driver and vehicle factors that might have contributed to the cause of the crash, and severity of the crash.

2.1. Scenario Coding Schemes

Appendix A presents coding schemes to identify common pre-crash scenarios leading to all single-vehicle and multi-vehicle (≥ 2) crashes based on GES variables and codes. A total of 46 pre-crash scenarios are listed in a selected order starting with scenarios associated with crash contributing factors such as vehicle control loss and driver violation of red light/stop sign (numbers 2-6). Such scenarios result in different crash types. For example, loss of vehicle control due to excessive speed could lead to a vehicle running off the road, rear-ending another vehicle in front of it, or encroaching into another lane and side-swiping an adjacent vehicle. From a crash avoidance perspective, the problem of vehicle control loss is identical in all three cases. A potential crash countermeasure function would detect the excessive speed or the imminent loss of control regardless of what crash type these conditions might lead to. Therefore, scenarios based on crash contributing factors in Appendix A supersede remaining scenarios that represent dynamically distinct driving situations based on vehicle movements and dynamic states. The new pre-crash scenario typology was then created by deducting the scenarios in the same order listed in Appendix A using the process of elimination. The sum of the resulting frequency distribution adds to 100 percent, and thus eliminating double counting of crashes in each of the scenarios.

The Accident Type, Movement Prior to Critical Event, and Critical Event variables from the GES Vehicle File were primarily used to identify dynamically distinct pre-crash scenarios. The first event in a crash from the GES Event File helped to distinguish pre-crash scenarios in multi-vehicle crashes. In addition to these variables, the coding schemes utilize the following GES variables:

- Traffic Control Device: Indicates whether or not a traffic control device was present for the crash and the type of traffic control device.
- Violations Charged: Indicates which violations are cited to drivers.
- First Harmful Event: Indicates the first property damaging or injury-producing event in the crash.

- Crash Event Sequence Number: Number assigned to each harmful event in a crash, in chronological order.
- Vehicle Number-This Vehicle: Number assigned to an in-transport motor vehicle involved in the event.
- Vehicle Number-Other Vehicle or Object Contacted: Vehicle number of the other vehicle or object hit, or the type of non-collision involved in the event.
- Vehicle Role: Indicates vehicle role (e.g., striking, struck) in single or multi-vehicle crashes.
- Rollover Type: Indicates if a rollover occurred (tripped or untripped). Rollover is defined as any vehicle rotation of 90 degrees or more about any true longitudinal or lateral axis. Rollover can occur at any time during the crash.
- Hit-and-run: It is coded when a motor vehicle in transport, or its driver, departs from the scene; vehicles not in transport are excluded. It does not matter whether the hit-and-run vehicle was striking or struck.
- Number of Vehicles Involved: Indicates the number of vehicles involved in the crash.

The following GES variables and codes were queried to identify the light vehicle:

- Body Type (Hot-Deck Imputed) = 01 – 22, 28 – 41, and 45 – 49
- Special Use = 00. This variable indicates whether the vehicle has a special use, meaning "in use" and not necessarily emergency use.

2.2. Crash Contributing Factors and Circumstances

Statistical description of crash contributing factors and circumstances was performed for each of the pre-crash scenarios that made up the final list of all scenarios leading to light-vehicle crashes. These factors and circumstances were broken down into three categories: driving environment, driver, and vehicle.

The following GES variables describe the driving environment:

- Light Condition: General light conditions at the time of the crash, including light from external roadway illumination fixtures.
- Atmospheric Conditions: General atmospheric conditions at the time of the crash (e.g., no adverse conditions, rain, sleet, fog, etc.).
- Roadway Surface Condition: Condition of road surface at the time of the crash (e.g., dry, wet, ice, etc.).
- Roadway Alignment: Horizontal alignment of roadway (straight or curve).
- Roadway Profile: Vertical alignment of roadway (e.g., level, grade etc.).
- Land Use: Population of the area associated with the police jurisdiction from which the crash report is selected. An area is considered rural if its population is less than or equal to 50,000.
- Day of Week
- Relation to Roadway: Indicates the location of the first harmful event.

- Relation to Junction: Indicates if the first harmful event is located within a junction or interchange area. If the first harmful event occurs off the roadway, the location classified is the point of departure.
- Posted Speed Limit
- Traffic Control Device

The following GES variables depict the driver factors:

- Driver Drinking in Vehicle: Reports alcohol use by driver of the vehicle.
- Driver's Vision Obscured by: Identifies visual circumstances that may have contributed to the cause of the crash.
- Driver Distracted by: Identifies a distraction that may have influenced driver performance and contributed to the cause of the crash. The distraction can be either inside or outside the vehicle.
- Speed Related: Indicates whether speed is a contributing factor to the cause of the crash.
- Violations Charged
- Person's Physical Impairment: Identifies physical impairments (e.g., ill, drowsy, deaf, etc.) for all drivers, which may have contributed to the cause of the crash.
- Sex: Male or female
- Age: This report classifies younger drivers as age 24 or younger, middle-aged drivers as between the ages of 25 and 64, and older drivers as age 65 or older.

The following GES variables portray the vehicle factors:

- Vehicle Contributing Factors: Indicates vehicle factors that may have contributed to the cause of the crash (e.g., tires, brakes, wipers, etc.)
- Rollover Type
- Movement Prior to Critical Event: (This variable is listed here so as to help in identifying dynamic variations of already-defined pre-crash scenarios).
- Driver Maneuvered to Avoid: Identifies an action taken by the driver to avoid something or someone in the road. The maneuver may have subsequently contributed to the cause of the crash.
- Corrective Action Attempted: Describes the actions taken by the driver of the vehicle in response to the impending danger. Because this variable focuses upon the driver's action just prior to the first harmful event, it is coded independently of any maneuvers associated with this vehicle's Accident Type. It should be noted that this variable reports many unknowns as seen in the results presented in this report. This same variable in the Crashworthiness Data System crash database provides a better description of driver evasive maneuvers in response to the critical event.

2.3. Societal Harm Measures

This report determines the frequency of occurrence for each pre-crash scenario in the new typology. It also estimates for each scenario its concomitant societal harm expressed in terms of economic cost or functional years lost. The "functional years lost" measure was selected for this analysis over other measures such as "equivalent lives" in order to harmonize with automakers who have recently adopted this measure in their crash avoidance research [1, 2]. These harm measures are derived from the maximum injury severity of all people involved in a specific crash scenario.

Economic Cost

Economic costs in this report account for goods and services that must be purchased or productivity that is lost as a result of motor vehicle crashes. They do not represent the intangible consequences of these events to individuals and families, such as pain and suffering and loss of life. Economic costs of crashes include lost productivity, medical costs, legal and court costs, emergency service costs, insurance administration costs, travel delay, property damage, and workplace losses.

The economic cost of crashes is computed on the basis of injury severity to the occupants of each vehicle involved in the crash according to the Abbreviated Injury Scale (AIS). The AIS is a classification system for assessing impact injury severity developed by the Association for the Advancement of Automotive Medicine. It provides the basis for stratifying the economic costs of crashes by injury severity. The Maximum Abbreviated Injury Scale (MAIS) is a function of AIS on a single injured person that measures overall maximum injury severity. Significant elements of economic loss, such as medical costs and lost productivity, are highly dependent on injury outcome.

GES does not provide detailed information regarding injury severity based on the AIS coding scheme. Instead, GES records injury severity by crash victim on the KABCO scale from police crash reports. Police reports in almost every State use KABCO to classify crash victims as K – killed, A – incapacitating injury, B – non-incapacitating injury, C – possible injury, O – no apparent injury, or ISU – Injury Severity Unknown. The KABCO coding scheme allows non-medically trained persons to make on-scene injury assessments without a hands-on examination. However, KABCO ratings are imprecise and inconsistently coded between States and over time. To estimate injuries based on the MAIS coding structure, a translator derived from 1982–1986 NASS data was applied to the GES police-reported injury profile [5]. The following matrix equation shows the multiplicative factors used to convert injury severity from KABCO to MAIS designations:

$$\begin{bmatrix} MAIS0 \\ MAIS1 \\ MAIS2 \\ MAIS3 \\ MAIS4 \\ MAIS5 \\ MAIS6 \end{bmatrix} = \begin{bmatrix} 0 & 0.01516 & 0.04938 & 0.19919 & 0.92423 & 0.07523 \\ 0 & 0.49183 & 0.79229 & 0.71729 & 0.07342 & 0.70581 \\ 0 & 0.27920 & 0.12487 & 0.06761 & 0.00206 & 0.15708 \\ 0 & 0.16713 & 0.03009 & 0.01509 & 0.00029 & 0.04343 \\ 0 & 0.02907 & 0.00267 & 0.00064 & 0.00001 & 0.01712 \\ 0 & 0.01762 & 0.00069 & 0.00018 & 0.00000 & 0.00134 \\ 1 & 0 & 0 & 0 & 0 & 0 \end{bmatrix} \begin{bmatrix} K \\ A \\ B \\ C \\ O \\ ISU \end{bmatrix}$$

It should be noted that K injuries in KABCO are converted only to fatalities and non-K injuries in KABCO are converted to MAIS 0-5 injuries. NHTSA recommends that fatal crashes and fatalities be extracted from the Fatality Analysis Reporting System (FARS), not GES, since it contains records on all fatal traffic crashes and thus provides a more accurate representation of fatal crashes and fatalities than the sample contained in GES. This report, however, counts fatalities from GES because FARS does not contain the Accident Type and Critical Event variables needed to identify the pre-crash scenarios of the new typology.

Table 3 provides MAIS values based on the 2000 crash economic cost [6]. These values are assigned to occupants of crash-involved vehicles in which one or more person suffered an injury. An amount of $2,532 was allocated to each property-damage-only (PDO) vehicle, referring to a vehicle that was damaged in a crash but no occupant was injured. All PDO vehicles, including those involved in injury crashes, were counted under PDO vehicles. The total economic costs of motor vehicle crashes in 2000 were estimated at $230.6 billion. Estimates of the number of crashes that occurred in 2000 included police-reported crashes from the 2000 GES as well as a significant number of non-reported crashes.

Table 3. MAIS Levels and Unit Costs in 2000 Dollars

MAIS	Severity	2000 $
0	Uninjured	1,962
1	Minor	10,562
2	Moderate	66,820
3	Serious	186,097
4	Severe	348,133
5	Critical	1,096,161
6	Fatal	977,208

Functional Years Lost

Functional years lost is a non-monetary measure that sums the years of life lost to fatal injury and the years of functional capacity lost to nonfatal injury [7]. This measure does not mirror the monetary economic cost. It assigns a different value to the relative severity of injuries suffered from motor vehicle crashes. Table 4 presents the functional years lost by MAIS levels.

Table 4. Functional Years Lost by MAIS Per-Unit Basis

MAIS	Severity	Functional Years Lost
1	Minor	0.07
2	Moderate	1.1
3	Serious	6.5
4	Severe	16.5
5	Critical	33.1
6	Fatal	42.7

3. DESCRIPTION OF LIGHT-VEHICLE CRASHES

This section presents statistics on the frequency of occurrence, severity, and number of vehicles involved for light-vehicle police-reported crashes based on the 2004 GES. These statistics are also compared to those of all-vehicle crashes. In addition, this section describes driving environment, driver, and vehicle factors that may have contributed to the cause of light-vehicle crashes.

3.1. Crash Severity

Approximately 6,170,000 police-reported crashes of all vehicle types involving 10,945,000 vehicles occurred in the United States based on 2004 GES statistics. A total of 15,342,000 people were involved in these crashes. About 2,819,000, or 18.4 percent of involved people were injured. By comparison, approximately 5,942,000 police-reported crashes involved at least one light vehicle, which accounted for 96 percent of all crashes in 2004. A total of 10,695,000 vehicles and 15,027,000 people were involved in these light-vehicle crashes resulting in 2,737,000 injured people. Table 5 compares the ratios of people involved by maximum injury severity between light-vehicle crashes and all-vehicle crashes using the KABCO and AIS injury scales. The two crash sets have almost similar injury distributions. "Died Prior" listed in the KABCO injury scale is indicated in police reports if the person died prior to the crash as a result of natural causes (e.g., heart attack), disease, drug overdose, or alcohol poisoning.

Table 5. Injury Severity Comparison between Light-Vehicle and All-Vehicle Crashes

	Injury Severity	Light-Vehicle Crashes	All-Vehicle Crashes	Light/All
KABCO Injury Scale	None	0.8179	0.8163	1.00
	Possible	0.1092	0.1085	1.01
	Non-incapacitating	0.0482	0.0495	0.97
	Incapacitating	0.0192	0.0201	0.95
	Fatal	0.0018	0.0020	0.92
	Unknown	0.0037	0.0037	1.00
	Died prior	0.000025	0.000024	1.02
	Sum	1.0000	1.0000	
AIS Injury Scale	None	0.7806	0.7791	1.00
	Minor	0.1886	0.1894	1.00
	Moderate	0.0210	0.0214	0.98
	Serious	0.0067	0.0069	0.97
	Severe	0.0008	0.0009	0.97
	Critical	0.00040	0.00041	0.96
	Fatal	0.0018	0.0020	0.92
	Sum	1.0000	1.0000	
	Injured people per crash	0.555	0.549	1.01

3.2. Crash Breakdown by Number of Vehicles Involved Per Crash

Figure 1 breaks down light-vehicle crashes and all-vehicle crashes by the number of vehicles involved per crash. Table 6 shows that the crash severity in terms of people involved or injured people per crash is the same between light-vehicle and all-vehicle crashes by the three categories of number of vehicles involved per crash.

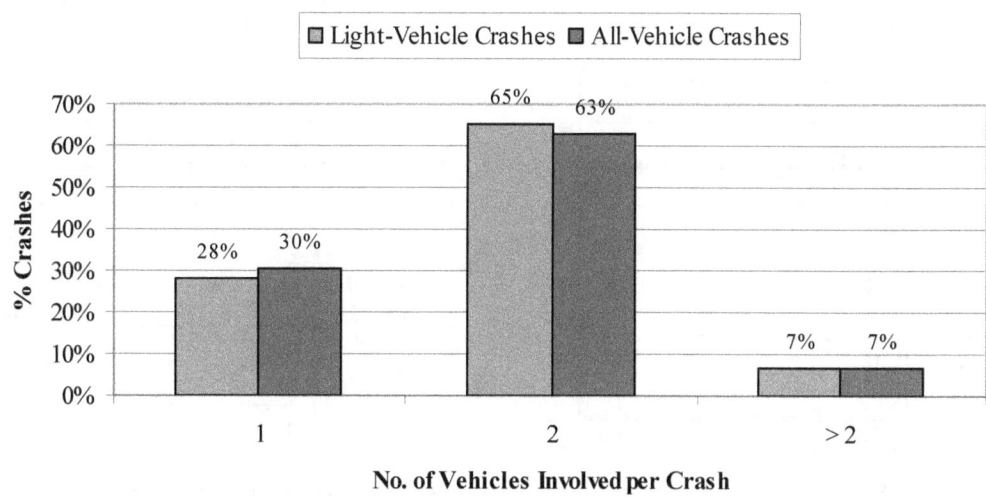

Figure 1. Distribution of Light-Vehicle and All-Vehicle Crashes by Number of Vehicles Involved per Crash

Table 6. Comparison of Crash Severity Between Light-Vehicle and All-Vehicle Crashes by Number of Vehicles Involved per Crash

	Type	Number of Crashes	Number of Persons	Number of Injured Persons	Persons per Crash	Injured Persons per Crash
1-Vehicle Crash	All	1,879,000	2,657,000	709,000	1.41	0.38
	Light	1,673,000	2,398,000	637,000	1.43	0.38
	Light/All	89.0%	90.3%	89.8%	1.01	1.01
2-Vehicle Crash	All	3,890,000	10,885,000	1,722,000	2.80	0.44
	Light	3,869,000	10,829,000	1,712,000	2.80	0.44
	Light/All	99.5%	99.5%	99.4%	1.00	1.00
Greater Than 2-Vehicle Crash	All	401,000	1,801,000	388,000	4.49	0.97
	Light	401,000	1,800,000	388,000	4.49	0.97
	Light/All	99.9%	99.9%	100.0%	1.00	1.00

3.3. Contributing Factors and Circumstances of Light-Vehicle Crashes

Table 7 presents statistics on driving environment factors, which are associated with all light-vehicle crashes.

Table 7. Driving Environment Statistics of Light-Vehicle Crashes

Category	Value	%		Category	Value	%
Lighting	Daylight	69%			Unknown	0.1%
	Dark Lighted	15%		Relation to Junction	Non-Junction	44%
	Dark	12%			Intersection	21.8%
	Dawn/Dusk	4%			Intersection-Related	19%
Weather	Clear	84%			Driveway/Alley	9.4%
	Adverse	16%			Entrance/Exit Ramp	3%
Road Surface	Dry	76%			Rail Grade Crossing	0.2%
	Wet/Slippery	24%			Other/Unknown	2%
Road Alignment	Straight	86%		Posted Speed Limit (mph)	<= 20	2%
	Curve	14%			25	13%
Road Profile	Level	78%			30	9%
	Other	22%			35	22%
Land Use	Rural	52%			40	9%
	Urban	48%			45	17%
Day	Weekday	77%			50	4%
	Weekend	23%			>= 55	24%
Relation to Roadway	On Roadway	78%		Traffic Control Device	No Traffic Controls	59%
	Shoulder/Parking Lane	4%			Traffic Signal	22%
	Off Roadway	17%			Stop/Yield Sign	12%
	Left Turn Lane	0.2%			Other	7%

A recent field operational test of a collision avoidance system, employing 66 subjects who drove instrumented vehicles as their own personal cars, revealed that approximately 10 percent and 25 percent of the distance traveled were done respectively in adverse weather and in the dark [8]. In addition, year 2000 data from the Bureau of Transportation Statistics showed that about 40 percent of the mileage driven in the United States was traveled in rural areas. Normalizing by distance traveled, light-vehicle crashes are over-represented at night, in adverse weather, and in rural areas. The reader is cautioned that this is a simple comparison of percentages and that these factors might not be over-represented.

Table 8 shows descriptive statistics of driver factors for light-vehicle crashes. Based on the 1995 Nationwide Personal Transportation Survey (NPTS), female drivers accounted for about 40 percent of the distance traveled by motor vehicles in the United States; younger and older drivers accumulated respectively 12 and 9 percent of the distance traveled [9]. Normalizing by distance traveled, younger drivers are greatly over-represented in light-vehicle crashes. As stated above, this over-representation of younger drivers is based on a simple comparison of percentages. Table 9 lists descriptive statistics of vehicle factors and evasive maneuvers for light-vehicle crashes.

Table 8. Driver Factors Statistics of All Light-Vehicle Drivers

Alcohol	Yes	4%
	No	96%
Vision Obscured	No Obstruction	71%
	Vision Obscured	3%
	Unknown	26%
Driver Distracted	Inattention	14%
	Sleepy	1%
	Not Distracted	44%
	Unknown	42%
Speeding	Yes	12%
	No	85%
	Unknown	3%
Violation	Speeding	0.1%
	Reckless	1%
	None	69%
	Other	27%
	Unknown	4%
Impairment	Ill/Blackout	0.2%
	Drowsy	1%
	None	93%
	Other	2%
	Unknown	4%
Gender	Male	56%
	Female	44%
Age	Younger <= 24	30%
	Middle = 25 to 64	63%
	Older >= 65	8%

Table 9. Vehicle Factor Statistics of All Light Vehicles

Contributing Factors	Yes	1%		Prior Corrective Action	0.3%
	No	91%		Other	1%
	Unknown	7%	**Driver Avoidance Maneuver**	Object in Road	0.2%
Rollover	Yes	3%		Poor Road Conditions	0.05%
	No	97%		Animal in Road	1%
Pre-Event Movement	No Driver Present	0.2%		Vehicle in Road	8%
	Going Straight	50%		Non-Motorist in Road	0.2%
	Decelerating in Traffic Lane	7%		Hit & Run	5%
	Accelerating in Traffic Lane	0.1%		No Driver Present	0.2%
	Starting in Traffic Lane	3%		Other Avoidance Maneuver	0.02%
	Stopped in Traffic Lane	14%		Unknown	56%
	Passing Another Vehicle	1%		None	29%
	Parked in Travel Lane	0.1%		Phantom Vehicle	0.2%
	Leaving a Parked Position	1%	**Corrective Action Attempted**	No Driver Present	0.2%
	Entering a Parked Position	0.2%		No Avoidance Maneuver	24%
	Turning Right	3%		Braking	6%
	Turning Left	10%		Releasing Brakes	0.01%
	Making U-turn	0.5%		Steering	4%
	Backing Up	2%		Braked and Steered	1%
	Negotiating a Curve	4%		Accelerated	0.2%
	Changing Lanes	3%		Accelerated and Steered	0.03%
	Merging	0.4%		Other	0.2%
				Unknown	65%

4. DETAILS OF NEW PRE-CRASH SCENARIO TYPOLOGY

The new pre-crash scenario typology of all light-vehicle crashes was derived by integrating lists of pre-crash scenarios from single-, two-, and multi-vehicle (more than two) crashes based on 2004 GES statistics. This section first presents results for each of the three crash categories. Afterward, the list of pre-crash scenarios for all light-vehicle crashes is discussed in terms of the frequency of occurrence, economic cost, and functional years lost. This is followed by a detailed description of crash characteristics for each scenario in the new pre-crash scenario typology. Such portrayal of scenario severity and crash characteristics will enable researchers to:

- Prioritize crash problem areas to be targeted for crash avoidance technology intervention
- Devise appropriate crash countermeasure concepts
- Determine applicable scenarios and define concomitant functional requirements
- Specify sensing and processing needs to assist drivers in preventing crashes via warning signals or automatic vehicle controls
- Develop guidelines for objective test procedures based on dynamic scenarios and driving characteristics most relevant to each applicable pre-crash scenario
- Estimate system effectiveness in each applicable pre-crash scenario and collectively assess potential safety benefits

This new typology is created to establish a consistent crash problem definition for developers of crash avoidance technologies, simplify crash characteristics for system designers, and prevent double counting of system safety benefits.

4.1. Single-Vehicle Pre-Crash Scenarios

Table 10 lists pre-crash scenarios of all single light-vehicle crashes in descending order in terms of frequency of occurrence. A total of 31 pre-crash scenarios represent 100 percent of all single light-vehicle crashes. The top three scenarios – control loss without prior vehicle action, road edge departure without prior vehicle maneuver, and animal crash without prior vehicle maneuver – account for about two thirds of all single light-vehicle crashes. The following twelve scenarios represent about 29 percent of all these crashes. The remaining sixteen pre-crash scenarios only correspond to five percent of all single light-vehicle crashes. It should be noted that vehicle action refers to a vehicle decelerating, accelerating, starting, passing, parking, turning, backing up, changing lanes, merging, and successful corrective action to a previous critical event. On the other hand, vehicle maneuver denotes passing, parking, turning, changing lanes, merging, and successful corrective action to a previous critical event.

Single light-vehicle crashes resulted in an estimated economic cost of about $37 billion and 1.1 million functional years lost. In terms of economic cost and functional years lost, the top three scenarios in descending order are:

1. Control loss without prior vehicle action: 36.7 percent of economic cost and 38.4 percent of functional years lost.
2. Road edge departure without prior vehicle maneuver: 24 percent of economic cost and 24.7 percent of functional years lost.
3. Pedestrian crash without prior vehicle maneuver: 10.3 percent of economic cost and 12.6 percent of functional years lost.

Thus, the top three scenarios listed above accounted for a total of 71 and 76 percent respectively of all economic cost and functional years lost due to single-vehicle light-vehicle crashes.

Table 10. Pre-Crash Scenarios of Single-Vehicle Light-Vehicle Crashes

No.	Scenario	Frequency	Rel. Freq.
1	Control Loss Without Prior Vehicle Action	471,000	28.15%
2	Road Edge Departure Without Prior Vehicle Maneuver	330,000	19.73%
3	Animal Crash Without Prior Vehicle Maneuver	300,000	17.91%
4	Control Loss With Prior Vehicle Action	74,000	4.41%
5	Road Edge Departure While Backing Up	66,000	3.93%
6	Road Edge Departure With Prior Vehicle Maneuver	66,000	3.92%
7	Object Crash Without Prior Vehicle Maneuver	55,000	3.26%
8	Pedestrian Crash Without Prior Vehicle Maneuver	37,000	2.22%
9	Vehicle Failure	33,000	1.99%
10	Object Crash With Prior Vehicle Maneuver	30,000	1.81%
11	Vehicle Changing Lanes – Same Direction	29,000	1.75%
12	Pedalcyclist Crash Without Prior Vehicle Maneuver	23,000	1.40%
13	Vehicle(s) Not Making a Maneuver – Opposite Direction	23,000	1.40%
14	Animal Crash With Prior Vehicle Maneuver	23,000	1.37%
15	Pedalcyclist Crash With Prior Vehicle Maneuver	18,000	1.07%
16	Non-Collision Incident	17,000	1.00%
17	Evasive Action Without Prior Vehicle Maneuver	16,000	0.98%
18	Pedestrian Crash With Prior Vehicle Maneuver	16,000	0.98%
19	Lead Vehicle Decelerating	9,000	0.55%
20	Vehicle(s) Turning at Non-Signalized Junctions	7,000	0.43%
21	Lead Vehicle Stopped	4,000	0.26%
22	Running Stop Sign	4,000	0.25%
23	No Driver Present	4,000	0.24%
24	Evasive Action With Prior Vehicle Maneuver	4,000	0.21%
25	On-Road Rollover	3,000	0.21%
26	Straight Crossing Paths at Non-Signalized Junctions	2,000	0.15%
27	Vehicle(s) Making a Maneuver – Opposite Direction	2,000	0.12%
28	Following Vehicle Making a Maneuver	2,000	0.12%
29	Lead Vehicle Moving at Lower Constant Speed	1,000	0.07%
30	Running Red Light	1,000	0.06%
31	Vehicle(s) Parking – Same Direction	1,000	0.05%

4.2. Two-Vehicle Pre-Crash Scenarios

Table 11 ranks pre-crash scenarios of two-vehicle crashes in descending order in terms of frequency of occurrence. A total of 31 pre-crash scenarios represent 99.3 percent of all two-vehicle crashes involving at least one light vehicle. The top three scenarios – lead vehicle stopped, vehicle(s) turning at non-signalized junctions, and lead vehicle decelerating – account for about 40 percent of all two-vehicle crashes and the following five scenarios represent about 31 percent of all these crashes. The remaining 23 pre-crash scenarios correspond to 28 percent of these crashes. There are "other" scenarios that only account for 0.7 percent of two-vehicle crashes involving at least one light vehicle including animal and cyclist with prior vehicle maneuver (0.01 percent each), on-road rollover (0.01 percent), hit-and-run (0.13 percent), and other non-specific or no-details scenarios. In about 50 percent of the lead-vehicle-stopped crashes, the lead vehicle first decelerates to a stop and is later struck by the following vehicle. This typically happens in the presence of a traffic control device or the lead vehicle is slowing down to make a turn. Thus, this particular scenario overlaps with the lead-vehicle-decelerating scenario.

Two-vehicle crashes involving at least one light vehicle resulted in an estimated economic cost of about $69 billion and 1.4 million functional years lost. In terms of economic cost, the top three scenarios in descending order are:

1. Lead vehicle stopped (14.9%)
2. Vehicle(s) turning at non-signalized junctions (10%)
3. Straight crossing paths at non-signalized junctions (9.9%)

The top three scenarios listed above accounted for a total of 34.9 percent of all economic cost due to two-vehicle light-vehicle crashes. In terms of functional years lost, the top three scenarios in descending order are:

1. Straight crossing paths at non-signalized junctions (11.6%)
2. Opposite direction without prior vehicle maneuver (11.6%)
3. Lead vehicle stopped (10.9%)

The top three scenarios listed above resulted in a total of 34 percent of all functional years lost due to two-vehicle light-vehicle crashes.

Table 11. Pre-Crash Scenarios of Two-Vehicle Light-Vehicle Crashes

No.	Scenario	Frequency	Rel. Freq.
1	Lead Vehicle Stopped	792,000	20.46%
2	Vehicle(s) Turning at Non-Signalized Junctions	419,000	10.83%
3	Lead Vehicle Decelerating	347,000	8.96%
4	Vehicle(s) Changing Lanes – Same Direction	295,000	7.62%
5	Straight Crossing Paths at Non-Signalized Junctions	252,000	6.52%
6	Running Red Light	233,000	6.02%
7	Vehicle(s) Turning – Same Direction	220,000	5.68%
8	LTAP/OD at Signalized Junctions	205,000	5.29%
9	Lead Vehicle Moving at Lower Constant Speed	186,000	4.82%
10	LTAP/OD at Non-Signalized Junctions	181,000	4.68%
11	Backing Up Into Another Vehicle	131,000	3.38%
12	Vehicle(s) Not Making a Maneuver – Opposite Direction	94,000	2.43%
13	Vehicle(s) Drifting – Same Direction	91,000	2.35%
14	Following Vehicle Making a Maneuver	74,000	1.92%
15	Control Loss Without Prior Vehicle Action	52,000	1.33%
16	Vehicle(s) Parking – Same Direction	47,000	1.21%
17	Running Stop Sign	43,000	1.12%
18	Evasive Action Without Prior Vehicle Maneuver	37,000	0.95%
19	Vehicle Turning Right at Signalized Junctions	34,000	0.89%
20	Control Loss With Prior Vehicle Action	26,000	0.68%
21	Non-Collision Incident	25,000	0.64%
22	Lead Vehicle Accelerating	16,000	0.41%
23	Vehicle(s) Making a Maneuver – Opposite Direction	13,000	0.33%
24	Evasive Action With Prior Vehicle Maneuver	8,000	0.21%
25	Vehicle Failure	8,000	0.20%
26	Animal Crash Without Prior Vehicle Maneuver	6,000	0.14%
27	Road Edge Departure Without Prior Vehicle Maneuver	3,000	0.08%
28	Pedestrian Crash Without Prior Vehicle Maneuver	2,000	0.05%
29	Road Edge Departure With Prior Vehicle Maneuver	2,000	0.04%
30	Pedestrian Crash With Prior Vehicle Maneuver	1,000	0.02%
31	Pedalcyclist Crash Without Prior Vehicle Maneuver	1,000	0.02%
32	Other	28,000	0.73%

4.3. Multi-Vehicle Pre-Crash Scenarios

Table 12 ranks pre-crash scenarios of multi-vehicle (more than two) crashes involving at least one light vehicle in descending order in terms of frequency of occurrence. A total of 24 pre-crash scenarios represent 99.4 percent of all these crashes. The top three scenarios – lead vehicle stopped, decelerating, and moving at lower constant speed – account for 68 percent of all multi-vehicle crashes and lead mostly to rear-end crashes. The following 11 scenarios represent about 27 percent of all these crashes. The remaining 10 pre-crash scenarios correspond to only 4 percent of these crashes. There are "other" scenarios that only account for 0.6 percent of multi-vehicle crashes involving at least one light vehicle including road edge departure with prior vehicle maneuver, animal and pedestrian without prior vehicle maneuver, backing up into another vehicle, parking, on-road rollover, hit-and-run, and other non-specific or no-details scenarios.

Multi-vehicle light-vehicle crashes resulted in an estimated economic cost of about $14 billion and 292 thousand functional years lost based on 2004 GES statistics. The top three scenarios, accounting for a total of 57.8 percent of all direct economic cost, are listed below in descending order:

1. Lead vehicle stopped (35.9%)
2. Lead vehicle decelerating (14.8%)
3. Opposite direction without prior vehicle maneuver (7.1%)

The top three scenarios, resulting in a total of 55 percent of all functional years lost, are listed below in descending order:

1. Lead vehicle stopped (29.6%)
2. Lead vehicle decelerating (13.3%)
3. Opposite direction without prior vehicle maneuver (11.7%)

Table 12. Pre-Crash Scenarios of Multi-Vehicle Light-Vehicle Crashes

No.	Scenario	Frequency	Rel. Freq.
1	Lead Vehicle Stopped	179,000	44.56%
2	Lead Vehicle Decelerating	72,000	18.05%
3	Lead Vehicle Moving at Lower Constant Speed	22,000	5.50%
4	Running Red Light	20,000	4.93%
5	LTAP/OD at Signalized Junctions	16,000	3.91%
6	Vehicle(s) Changing Lanes – Same Direction	14,000	3.54%
7	Following Vehicle Making a Maneuver	9,000	2.25%
8	Straight Crossing Paths at Non-Signalized Junctions	9,000	2.24%
9	LTAP/OD at Non-Signalized Junctions	9,000	2.23%
10	Vehicle(s) Turning at Non-Signalized Junctions	9,000	2.14%
11	Vehicle(s) Drifting – Same Direction	7,000	1.81%
12	Control Loss Without Prior Vehicle Action	6,000	1.62%
13	Vehicle(s) Not Making a Maneuver – Opposite Direction	6,000	1.60%
14	Non-Collision Incident	5,000	1.15%
15	Evasive Action Without Prior Vehicle Maneuver	3,000	0.77%
16	Lead Vehicle Accelerating	3,000	0.67%
17	Control Loss With Prior Vehicle Action	3,000	0.64%
18	Vehicle(s) Turning – Same Direction	2,000	0.49%
19	Evasive Action With Prior Vehicle Maneuver	1,000	0.33%
20	Vehicle Failure	1,000	0.32%

No.	Scenario	Frequency	Rel. Freq.
21	Running Stop Sign	1,000	0.24%
22	Vehicle Turning Right at Signalized Junctions	1,000	0.17%
23	Vehicle(s) Making a Maneuver – Opposite Direction	1,000	0.16%
24	Road Edge Departure Without Prior Vehicle Maneuver	1,000	0.14%
25	Other	2,000	0.55%

4.4. All Light-Vehicle Pre-Crash Scenarios

Table 13 ranks pre-crash scenarios of all light-vehicle crashes in descending order in terms of frequency of occurrence. A total of 36 pre-crash scenarios represent 99.4 percent of all light-vehicle crashes. The top scenario with an individual relative frequency over ten percent – lead vehicle stopped – accounts for 16 percent of all light-vehicle crashes. The following six scenarios with an individual relative frequency between 5 and 10 percent represent about 40 percent of all these crashes. The remaining 29 pre-crash scenarios correspond to 43 percent of all light-vehicle crashes. There are "other" scenarios that only account for 0.6 percent of all light-vehicle crashes including on-road rollover (0.06%), hit-and-run (0.09%), no driver present (0.07%), and other non-specific or no-details scenarios.

Table 14 ranks pre-crash scenarios of all light-vehicle crashes in descending order in terms of economic cost. Overall, police-reported light-vehicle crashes resulted in an estimated cost of $120 billion based on 2004 GES statistics. It should be noted that these societal harm estimates are based solely on police-reported crashes captured by the GES crash database, excluding a large number of non-police-reported crashes. The top three scenarios – control loss without prior vehicle action, lead vehicle stopped, and road edge departure without prior vehicle maneuver – account for a total of 34 percent of all economic cost.

Table 15 ranks pre-crash scenarios of all light-vehicle crashes in descending order in terms of functional years lost, which totaled about 2,767,000 years based on 2004 GES statistics. The top five scenarios, accounting for a total of 49 percent of all functional years lost, are listed below in descending order along with their respective ranks in terms of frequency of occurrence (frequency) and economic cost (cost):

1. Control loss without prior vehicle action – second in frequency and first in cost
2. Road edge departure without prior vehicle maneuver – fifth in frequency and third in cost
3. Lead vehicle stopped – first in frequency and second in cost
4. Opposite direction without prior vehicle maneuver – fifteenth in frequency and seventh in cost
5. Straight crossing paths at non-signalized junctions – eighth in frequency and fifth in cost

The following lists three scenarios that appear in the top five pre-crash scenarios in frequency of occurrence, economic cost, and functional years lost:

1. Control loss without prior vehicle action
2. Lead vehicle stopped
3. Road edge departure without prior vehicle maneuver

Table 13. Pre-Crash Scenarios of All Light-Vehicle Crashes

No.	Scenario	1-Frequency	Frequency	Rel. Freq.
1	Lead Vehicle Stopped	974,855	975,000	16.41%
2	Control Loss Without Prior Vehicle Action	528,930	529,000	8.90%
3	Vehicle(s) Turning at Non-Signalized Junctions	434,892	435,000	7.32%
4	Lead Vehicle Decelerating	428,067	428,000	7.20%
5	Road Edge Departure Without Prior Vehicle Maneuver	333,706	334,000	5.62%
6	Vehicle(s) Changing Lanes – Same Direction	338,309	338,000	5.69%
7	Animal Crash Without Prior Vehicle Maneuver	305,102	305,000	5.13%
8	Straight Crossing Paths at Non-Signalized Junctions	263,840	264,000	4.44%
9	Running Red Light	253,618	254,000	4.27%
10	Vehicle(s) Turning – Same Direction	221,791	222,000	3.73%
11	LTAP/OD at Signalized Junctions	220,206	220,000	3.71%
12	Lead Vehicle Moving at Lower Constant Speed	209,610	210,000	3.53%
13	LTAP/OD at Non-Signalized Junctions	189,816	190,000	3.19%
14	Backing Up Into Another Vehicle	130,701	131,000	2.20%
15	Vehicle(s) Not Making a Maneuver – Opposite Direction	123,699	124,000	2.08%
16	Control Loss With Prior Vehicle Action	102,617	103,000	1.73%
17	Vehicle(s) Drifting – Same Direction	97,973	98,000	1.65%
18	Following Vehicle Making a Maneuver	85,373	85,000	1.44%
19	Road Edge Departure With Prior Vehicle Maneuver	67,528	68,000	1.14%
20	Road Edge Departure While Backing Up	65,809	66,000	1.11%
21	Object Crash Without Prior Vehicle Maneuver	54,526	55,000	0.92%
22	Evasive Action Without Prior Vehicle Maneuver	56,199	56,000	0.95%
23	Vehicle(s) Parking – Same Direction	48,138	48,000	0.81%
24	Running Stop Sign	48,296	48,000	0.81%
25	Non-Collision Incident	45,910	46,000	0.77%
26	Vehicle Failure	42,147	42,000	0.71%
27	Pedestrian Crash Without Prior Vehicle Maneuver	39,324	39,000	0.66%
28	Vehicle Turning Right at Signalized Junctions	34,951	35,000	0.59%
29	Object Crash With Prior Vehicle Maneuver	30,301	30,000	0.51%
30	Pedalcyclist Crash Without Prior Vehicle Maneuver	24,071	24,000	0.41%
31	Animal Crash With Prior Vehicle Maneuver	23,322	23,000	0.39%
32	Pedalcyclist Crash With Prior Vehicle Maneuver	18,325	18,000	0.31%
33	Pedestrian Crash With Prior Vehicle Maneuver	17,118	17,000	0.29%
34	Lead Vehicle Accelerating	18,722	19,000	0.32%
35	Vehicle(s) Making a Maneuver – Opposite Direction	15,472	15,000	0.26%
36	Evasive Action With Prior Vehicle Maneuver	13,120	13,000	0.22%
37	Other	35,859	36,000	0.60%

Table 14. Ranking of Light-Vehicle Pre-Crash Scenarios by Economic Cost

No.	Scenario	Cost ($)	Rel. Cost
1	Control Loss Without Prior Vehicle Action	$ 15,796,000,000	13.18%
2	Lead Vehicle Stopped	$ 15,388,000,000	12.84%
3	Road Edge Departure Without Prior Vehicle Maneuver	$ 9,005,000,000	7.51%
4	Vehicle(s) Turning at Non-Signalized Junctions	$ 7,343,000,000	6.13%
5	Straight Crossing Paths at Non-Signalized Junctions	$ 7,290,000,000	6.08%
6	Running Red Light	$ 6,627,000,000	5.53%
7	Vehicle(s) Not Making a Maneuver - Opposite Direction	$ 6,407,000,000	5.35%
8	Lead Vehicle Decelerating	$ 6,390,000,000	5.33%
9	LTAP/OD at Signalized Junctions	$ 5,749,000,000	4.80%
10	LTAP/OD at Non-Signalized Junctions	$ 5,137,000,000	4.29%
11	Vehicle(s) Changing Lanes - Same Direction	$ 4,247,000,000	3.54%
12	Pedestrian Crash Without Prior Vehicle Maneuver	$ 4,022,000,000	3.36%
13	Lead Vehicle Moving at Lower Constant Speed	$ 3,910,000,000	3.26%
14	Vehicle(s) Turning - Same Direction	$ 2,810,000,000	2.34%
15	Control Loss With Prior Vehicle Action	$ 1,970,000,000	1.64%
16	Animal Crash Without Prior Vehicle Maneuver	$ 1,632,000,000	1.36%
17	Vehicle(s) Drifting - Same Direction	$ 1,383,000,000	1.15%
18	Evasive Action Without Prior Vehicle Maneuver	$ 1,349,000,000	1.13%
19	Running Stop Sign	$ 1,310,000,000	1.09%
20	Pedalcyclist Crash Without Prior Vehicle Maneuver	$ 1,301,000,000	1.09%
21	Following Vehicle Making a Maneuver	$ 1,212,000,000	1.01%
22	Road Edge Departure With Prior Vehicle Maneuver	$ 1,144,000,000	0.95%
23	Vehicle Failure	$ 1,051,000,000	0.88%
24	Backing Up Into Another Vehicle	$ 947,000,000	0.79%
25	Vehicle(s) Making a Maneuver - Opposite Direction	$ 943,000,000	0.79%
26	Pedestrian Crash With Prior Vehicle Maneuver	$ 843,000,000	0.70%
27	Object Crash Without Prior Vehicle Maneuver	$ 687,000,000	0.57%
28	Vehicle(s) Parking - Same Direction	$ 623,000,000	0.52%
29	Non-Collision Incident	$ 592,000,000	0.49%
30	Pedalcyclist Crash With Prior Vehicle Maneuver	$ 523,000,000	0.44%
31	Vehicle Turning Right at Signalized Junctions	$ 355,000,000	0.30%
32	Road Edge Departure While Backing Up	$ 350,000,000	0.29%
33	Lead Vehicle Accelerating	$ 273,000,000	0.23%
34	Evasive Action With Prior Vehicle Maneuver	$ 198,000,000	0.17%
35	Object Crash With Prior Vehicle Maneuver	$ 155,000,000	0.13%
36	Animal Crash With Prior Vehicle Maneuver	$ 120,000,000	0.10%
37	Other	$ 764,000,000	0.64%

Table 15. Ranking of Light-Vehicle Pre-Crash Scenarios by Functional Years Lost

No.	Scenario	Years Lost	Rel. Yrs Lost
1	Control Loss Without Prior Vehicle Action	478,000	17.27%
2	Road Edge Departure Without Prior Vehicle Maneuver	270,000	9.76%
3	Lead Vehicle Stopped	240,000	8.69%
4	Vehicle(s) Not Making a Maneuver – Opposite Direction	206,000	7.44%
5	Straight Crossing Paths at Non-Signalized Junctions	174,000	6.29%
6	Pedestrian Crash Without Prior Vehicle Maneuver	144,000	5.21%
7	Vehicle(s) Turning at Non-Signalized Junctions	138,000	5.00%
8	Running Red Light	135,000	4.87%
9	LTAP/OD at Signalized Junctions	121,000	4.36%
10	LTAP/OD at Non-Signalized Junctions	113,000	4.09%
11	Lead Vehicle Decelerating	100,000	3.62%
12	Lead Vehicle Moving at Lower Constant Speed	78,000	2.81%
13	Vehicle(s) Changing Lanes – Same Direction	71,000	2.57%
14	Control Loss With Prior Vehicle Action	49,000	1.76%
15	Vehicle(s) Turning – Same Direction	47,000	1.68%
16	Pedalcyclist Crash Without Prior Vehicle Maneuver	39,000	1.42%
17	Vehicle(s) Drifting – Same Direction	37,000	1.32%
18	Evasive Action Without Prior Vehicle Maneuver	36,000	1.31%
19	Road Edge Departure With Prior Vehicle Maneuver	34,000	1.22%
20	Vehicle(s) Making a Maneuver – Opposite Direction	32,000	1.14%
21	Running Stop Sign	28,000	1.02%
22	Vehicle Failure	26,000	0.93%
23	Pedestrian Crash With Prior Vehicle Maneuver	24,000	0.88%
24	Animal Crash Without Prior Vehicle Maneuver	24,000	0.86%
25	Object Crash Without Prior Vehicle Maneuver	19,000	0.68%
26	Following Vehicle Making a Maneuver	18,000	0.67%
27	Non-Collision Incident	13,000	0.45%
28	Vehicle(s) Parking – Same Direction	11,000	0.41%
29	Pedalcyclist Crash With Prior Vehicle Maneuver	11,000	0.39%
30	Backing Up Into Another Vehicle	9,000	0.32%
31	Road Edge Departure While Backing Up	6,000	0.21%
32	Lead Vehicle Accelerating	4,000	0.15%
33	Vehicle Turning Right at Signalized Junctions	4,000	0.15%
34	Evasive Action With Prior Vehicle Maneuver	4,000	0.13%
35	Object Crash With Prior Vehicle Maneuver	3,000	0.10%
36	Animal Crash With Prior Vehicle Maneuver	2,000	0.06%
37	Other	21,000	0.75%

4.5. Statistical Description of All Light-Vehicle Pre-Crash Scenarios

The following provides a detailed description for each of the 37 scenarios based on the same order as listed in Appendix A. Appendix B also lists in tabular format descriptive statistics about driving environment, driver, and vehicle factors for each of these scenarios.

Vehicle Failure

Typical Scenario: Vehicle is going straight in a rural area, in daylight, under clear weather conditions, on a dry road with a posted speed limit of 55 mph or more, and then loses control due to catastrophic component failure at a non-junction and runs off the road. Failure of tires, brakes, power train, steering system, and wheels contributed to about 95 percent of these crashes, with tires alone accounting for 62 percent of vehicle failure crashes.

Factor Over-Representation: Rural area, non-junction, high-speed road, younger driver, and rollover are over-represented (based on a simple comparison of percentages).

Dynamic Variations: Vehicle is negotiating a curve and then loses control due to component failure (24% of crashes).

Scenario Severity: Table below quantifies the annual severity of this crash scenario in terms of five different metrics based on 2004 GES statistics. This table also provides the ratios of people involved by maximum injury severity using the KABCO and AIS injury scales. About 1.78 percent of all people involved in this crash scenario suffered high-level MAIS 3+ injuries (serious, severe, critical, or fatal).

	Crash Severity	**Scenario**	**Scenario/All**
	No. of crashes	42,000	0.71%
	No. of vehicles involved	53,000	0.50%
	No. of people involved	89,000	0.59%
Societal Cost	Economic cost	$1,051,000,000	0.88%
	Functional years lost	26,000	0.93%
KABCO Injury Scale	None	0.718	0.878
	Possible	0.097	0.884
	Non-incapacitating	0.133	2.759
	Incapacitating	0.043	2.261
	Fatal	0.002	1.101
	Unknown	0.007	1.860
	Died prior	-	-
AIS Injury Scale	None	0.691	0.885
	Minor	0.253	1.344
	Moderate	0.038	1.796
	Serious	0.013	1.969
	Severe	0.002	2.175
	Critical	0.001	2.226
	Fatal	0.002	1.092
	Injured people per crash	0.655	1.181

Control Loss With Prior Vehicle Action

Typical Scenario: Vehicle is turning left or right at an intersection-related area, in daylight, under clear weather conditions, with a posted speed limit of 45 mph or less, and then loses control due to wet or slippery roads and runs off the road.

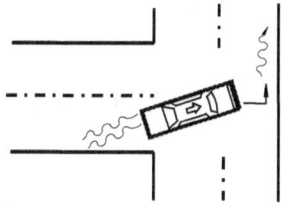

Factor Over-Representation: Dark, adverse weather, wet or slippery road, intersection-related, speeding, younger driver, and rollover are over-represented (based on a simple comparison of percentages).

Dynamic Variations: Vehicle is decelerating in the traffic lane or changing lanes and then loses control.

Scenario Severity: The table below quantifies the annual severity of this crash scenario in terms of five different metrics based on 2004 GES statistics. This table also provides the ratios of people involved by maximum injury severity using the KABCO and AIS injury scales. About 1.43 percent of all people involved in this crash scenario suffered high-level MAIS 3+ injuries (serious, severe, critical, or fatal).

	Crash Severity	**Scenario**	**Scenario/All**
	No. of crashes	103,000	1.73%
	No. of vehicles involved	135,000	1.26%
	No. of people involved	192,000	1.28%
Societal Cost	Economic cost	$1,970,000,000	1.64%
	Functional years lost	49,000	1.76%
KABCO Injury Scale	None	0.780	0.954
	Possible	0.105	0.959
	Non-incapacitating	0.074	1.532
	Incapacitating	0.035	1.827
	Fatal	0.002	1.263
	Unknown	0.004	0.999
	Died prior	-	-
AIS Injury Scale	None	0.747	0.956
	Minor	0.211	1.117
	Moderate	0.028	1.344
	Serious	0.010	1.497
	Severe	0.001	1.634
	Critical	0.001	1.746
	Fatal	0.002	1.252
	Injured people per crash	0.474	0.855

Control Loss Without Prior Vehicle Action

Typical Scenario: Vehicle is going straight in a rural area, in daylight, under adverse weather conditions, with a posted speed limit of 55 mph or more, and then loses control due to wet or slippery roads and runs off the road.

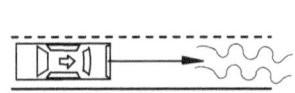

Factor Over-Representation: Dark, adverse weather, wet/slippery road, rural area, non-junction, high-speed road, speeding, younger driver, and rollover are over-represented (based on a simple comparison of percentages).

Dynamic Variations: Vehicle is negotiating a curve and loses control (42% of crashes).

Scenario Severity: Table below quantifies the annual severity of this crash scenario in terms of five different metrics based on 2004 GES statistics. This table also provides the ratios of people involved by maximum injury severity using the KABCO and AIS injury scales. About 2.67 percent of all people involved in this crash scenario suffered high-level MAIS 3+ injuries (serious, severe, critical, or fatal). Approximately 1,000 pedestrians were involved in this crash scenario.

	Crash Severity	**Scenario**	**Scenario/All**
	No. of crashes	529,000	8.90%
	No. of vehicles involved	596,000	5.57%
	No. of people involved	825,000	5.49%
Societal Cost	Economic cost	$15,796,000,000	13.18%
	Functional years lost	478,000	17.27%
KABCO Injury Scale	None	0.672	0.821
	Possible	0.139	1.271
	Non-incapacitating	0.121	2.506
	Incapacitating	0.056	2.928
	Fatal	0.008	4.443
	Unknown	0.004	1.163
	Died prior	0.0003	11.118
AIS Injury Scale	None	0.656	0.840
	Minor	0.275	1.459
	Moderate	0.042	2.006
	Serious	0.015	2.310
	Severe	0.002	2.565
	Critical	0.001	2.785
	Fatal	0.008	4.405
	Injured people per crash	0.537	0.967

Running Red Light

Typical Scenario: Vehicle is going straight in an urban area, in daylight, under clear weather conditions, with a posted speed limit of 35 mph; vehicle then runs a red light, crossing an intersection and colliding with another vehicle crossing the intersection from a lateral direction.

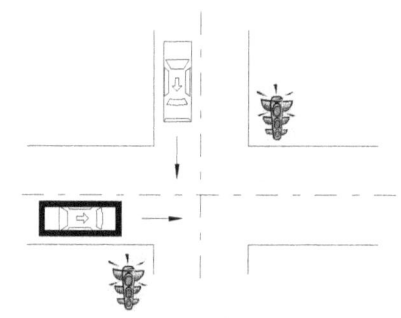

Factor Over-Representation: Urban area, inattention, female driver, and younger and older drivers are over-represented (based on a simple comparison of percentages).

Dynamic Variations: Vehicle runs a red light while turning left and collides with another straight crossing vehicle from a lateral direction.

Scenario Severity: Table below quantifies the annual severity of this crash scenario in terms of five different metrics based on 2004 GES statistics. This table also provides the ratios of people involved by maximum injury severity using the KABCO and AIS injury scales. About 1.18 percent of all people involved in this crash scenario suffered high-level MAIS 3+ injuries (serious, severe, critical, or fatal).

	Crash Severity	Scenario	Scenario/All
	No. of crashes	254,000	4.27%
	No. of vehicles involved	528,000	4.94%
	No. of people involved	740,000	4.92%
Societal Cost	Economic cost	$6,627,000,000	5.53%
	Functional years lost	135,000	4.87%
KABCO Injury Scale	None	0.726	0.888
	Possible	0.169	1.546
	Non-incapacitating	0.073	1.522
	Incapacitating	0.025	1.283
	Fatal	0.001	0.457
	Unknown	0.006	1.666
	Died prior	-	-
AIS Injury Scale	None	0.709	0.909
	Minor	0.249	1.320
	Moderate	0.030	1.422
	Serious	0.009	1.393
	Severe	0.001	1.366
	Critical	0.001	1.319
	Fatal	0.001	0.453
	Injured people per crash	0.848	1.528

Running Stop Sign

Typical Scenario: Vehicle is going straight in a rural area, in daylight, under clear weather conditions, with a posted speed limit of 35 mph or less; and runs a stop sign at an intersection.

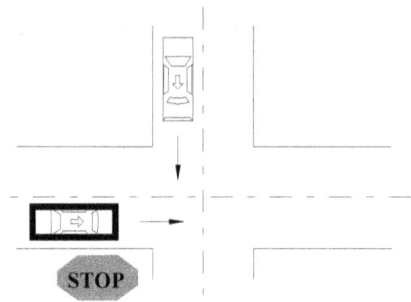

Factor Over-Representation: Low posted speed limit (35 mph or less), inattention, and younger and older drivers are over-represented (based on a simple comparison of percentages).

Dynamic Variations: Vehicle runs a stop sign while turning either left or right.

Scenario Severity: Table below quantifies the annual severity of this crash scenario in terms of five different metrics based on 2004 GES statistics. This table also provides the ratios of people involved by maximum injury severity using the KABCO and AIS injury scales. About 1.33 percent of all people involved in this crash scenario suffered high-level MAIS 3+ injuries (serious, severe, critical, or fatal).

	Crash Severity	Scenario	Scenario/All
	No. of crashes	48,000	0.81%
	No. of vehicles involved	93,000	0.87%
	No. of people involved	133,000	0.88%
Societal Cost	Economic cost	$1,310,000,000	1.09%
	Functional years lost	28,000	1.02%
KABCO Injury Scale	None	0.710	0.868
	Possible	0.162	1.487
	Non-incapacitating	0.088	1.830
	Incapacitating	0.026	1.386
	Fatal	0.001	0.671
	Unknown	0.012	3.169
	Died prior	-	-
AIS Injury Scale	None	0.694	0.889
	Minor	0.260	1.377
	Moderate	0.033	1.555
	Serious	0.010	1.530
	Severe	0.001	1.592
	Critical	0.001	1.448
	Fatal	0.001	0.665
	Injured people per crash	0.839	1.513

Road Edge Departure With Prior Vehicle Maneuver

Typical Scenario: Vehicle is turning left/right at an intersection-related location, in a rural area at night, under clear weather conditions, with a posted speed limit of 25 mph; and then departs the edge of the road.

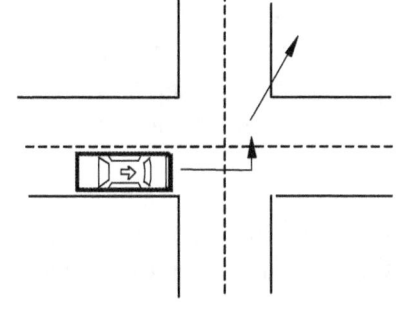

Factor Over-Representation: Dark, intersection-related, low-speed road, alcohol, inattention, and younger driver are over-represented (based on a simple comparison of percentages).

Dynamic Variations: Vehicle attempts to change lanes/pass or enter/leave a parking position and departs the edge of the road. The first harmful event of the "road edge departure with prior vehicle maneuver" scenario occurs at road shoulder or parking lane in one-third of these crashes. Moreover, the vehicle departs the road edge to the right in about two-thirds of these crashes.

Scenario Severity: Table below quantifies the annual severity of this crash scenario in terms of five different metrics based on 2004 GES statistics. This table also provides the ratios of people involved by maximum injury severity using the KABCO and AIS injury scales. About 1.42 percent of all people involved in this crash scenario suffered high-level MAIS 3+ injuries (serious, severe, critical, or fatal). Approximately 1,000 pedestrians were involved in this crash scenario.

	Crash Severity	**Scenario**	**Scenario/All**
	No. of crashes	68,000	1.14%
	No. of vehicles involved	70,000	0.65%
	No. of people involved	98,000	0.65%
Societal Cost	Economic cost	$1,144,000,000	0.95%
	Functional years lost	34,000	1.22%
KABCO Injury Scale	None	0.827	1.011
	Possible	0.059	0.540
	Non-incapacitating	0.079	1.642
	Incapacitating	0.022	1.130
	Fatal	0.005	2.925
	Unknown	0.008	2.162
	Died prior	-	-
AIS Injury Scale	None	0.781	1.000
	Minor	0.182	0.965
	Moderate	0.023	1.087
	Serious	0.007	1.115
	Severe	0.001	1.235
	Critical	0.0005	1.155
	Fatal	0.005	2.899
	Injured people per crash	0.318	0.574

Road Edge Departure Without Prior Vehicle Maneuver

Typical Scenario: Vehicle is going straight in a rural area at night, under clear weather conditions, with a posted speed limit of 55 mph or more, and departs the edge of the road at a non-junction area.

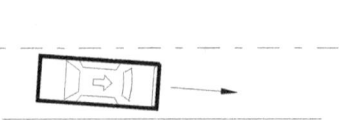

Factor Over-Representation: Dark, rural area, non-junction, alcohol, inattention, speeding, drowsiness, younger driver, and rollover are over-represented (based on a simple comparison of percentages).

Dynamic Variations: Vehicle is negotiating a curve and departs the edge of the road (26% of crashes). The first harmful event of the "road edge departure without prior vehicle maneuver" scenario occurs at road shoulder or parking lane in about 27 percent of these crashes. Moreover, the vehicle departs the road edge to the right in about two-thirds of these crashes.

Scenario Severity: Table below quantifies the annual severity of this crash scenario in terms of five different metrics based on 2004 GES statistics. This table also provides the ratios of people involved by maximum injury severity using the KABCO and AIS injury scales. About 2.79 percent of all people involved in this crash scenario suffered high-level MAIS 3+ injuries (serious, severe, critical, or fatal). Approximately 2,000 pedestrians were involved in this crash scenario.

	Crash Severity	**Scenario**	**Scenario/All**
	No. of crashes	334,000	5.62%
	No. of vehicles involved	338,000	3.16%
	No. of people involved	456,000	3.03%
Societal Cost	Economic cost	$9,005,000,000	7.51%
	Functional years lost	270,000	9.76%
KABCO Injury Scale	None	0.652	0.798
	Possible	0.131	1.201
	Non-incapacitating	0.141	2.930
	Incapacitating	0.058	3.023
	Fatal	0.008	4.410
	Unknown	0.009	2.572
	Died prior	-	-
AIS Injury Scale	None	0.638	0.817
	Minor	0.289	1.532
	Moderate	0.045	2.164
	Serious	0.016	2.462
	Severe	0.002	2.795
	Critical	0.001	2.915
	Fatal	0.008	4.371
	Injured people per crash	0.495	0.892

Road Edge Departure While Backing Up

Typical Scenario: Vehicle is backing up in an urban area, in daylight, under clear weather conditions, with a posted speed limit of 25 mph; and then departs the road edge on the shoulder/parking lane in a driveway/alley location.

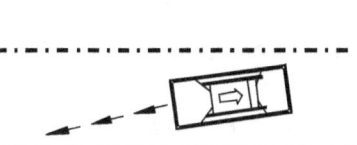

Factor Over-Representation: Driveway/alley location, low-speed road, alcohol, inattention, and younger driver are over-represented (based on a simple comparison of percentages).

Dynamic Variations: Vehicle is leaving/entering a parked position while backing up and departs the edge of the road.

Scenario Severity: Table below quantifies the annual severity of this crash scenario in terms of five different metrics based on 2004 GES statistics. This table also provides the ratios of people involved by maximum injury severity using the KABCO and AIS injury scales. About 0.27 percent of all people involved in this crash scenario suffered high-level MAIS 3+ injuries (serious, severe, critical, or fatal). Approximately 4,000 pedestrians were involved in this crash scenario.

	Crash Severity	Scenario	Scenario/All
	No. of crashes	66,000	1.11%
	No. of vehicles involved	66,000	0.62%
	No. of people involved	95,000	0.63%
Societal Cost	Economic cost	$350,000,000	0.29%
	Functional years lost	6,000	0.21%
KABCO Injury Scale	None	0.941	1.150
	Possible	0.037	0.342
	Non-incapacitating	0.016	0.336
	Incapacitating	0.003	0.131
	Fatal	0.001	0.358
	Unknown	0.002	0.605
	Died prior	-	-
AIS Injury Scale	None	0.878	1.125
	Minor	0.112	0.591
	Moderate	0.008	0.359
	Serious	0.002	0.275
	Severe	0.0002	0.227
	Critical	0.0001	0.165
	Fatal	0.001	0.355
	Injured people per crash	0.176	0.318

Animal Crash With Prior Vehicle Maneuver

Typical Scenario: Vehicle is leaving a parked position in a rural area at night, under clear weather conditions; and encounters an animal at a non-junction area.

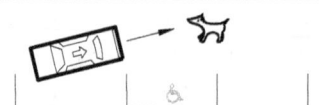

Factor Over-Representation: Dark, wet, or slippery road, rural area, non-junction, and high-speed road are over-represented (based on a simple comparison of percentages).

Dynamic Variations: Vehicle is passing another vehicle and encounters an animal.

Scenario Severity: Table below quantifies the annual severity of this crash scenario in terms of five different metrics based on 2004 GES statistics. This table also provides the ratios of people involved by maximum injury severity using the KABCO and AIS injury scales. About 0.36 percent of all people involved in this crash scenario suffered high-level MAIS 3+ injuries (serious, severe, critical, or fatal).

	Crash Severity	Scenario	Scenario/All
	No. of crashes	23,000	0.39%
	No. of vehicles involved	24,000	0.22%
	No. of people involved	27,000	0.18%
Societal Cost	Economic cost	$120,000,000	0.10%
	Functional years lost	2,000	0.06%
KABCO Injury Scale	None	0.889	1.087
	Possible	0.083	0.759
	Non-incapacitating	0.022	0.451
	Incapacitating	0.005	0.240
	Fatal	0.0002	0.099
	Unknown	0.002	0.498
	Died prior	-	-
AIS Injury Scale	None	0.839	1.075
	Minor	0.145	0.771
	Moderate	0.012	0.557
	Serious	0.003	0.449
	Severe	0.0003	0.344
	Critical	0.0001	0.286
	Fatal	0.0002	0.098
	Injured people per crash	0.186	0.336

Animal Crash Without Prior Vehicle Maneuver

Typical Scenario: Vehicle is going straight in a rural area at night, under clear weather conditions, with a posted speed limit of 55 mph or more; and encounters an animal at a non-junction location.

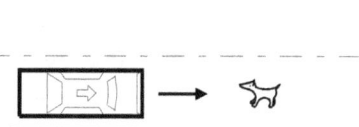

Factor Over-Representation: Dark, rural area, non-junction, and high-speed road are over-represented (based on a simple comparison of percentages).

Dynamic Variations: Vehicle is negotiating a curve and encounters an animal (11% of crashes).

Scenario Severity: Table below quantifies the annual severity of this crash scenario in terms of five different metrics based on 2004 GES statistics. This table also provides the ratios of people involved by maximum injury severity using the KABCO and AIS injury scales. About 0.38 percent of all people involved in this crash scenario suffered high-level MAIS 3+ injuries (serious, severe, critical, or fatal).

	Crash Severity	Scenario	Scenario/All
	No. of crashes	305,000	5.13%
	No. of vehicles involved	311,000	2.90%
	No. of people involved	414,000	2.76%
Societal Cost	Economic cost	$1,632,000,000	1.36%
	Functional years lost	24,000	0.86%
KABCO Injury Scale	None	0.921	1.126
	Possible	0.040	0.364
	Non-incapacitating	0.030	0.618
	Incapacitating	0.008	0.412
	Fatal	0.0001	0.065
	Unknown	0.001	0.324
	Died prior	-	-
AIS Injury Scale	None	0.861	1.103
	Minor	0.124	0.660
	Moderate	0.011	0.509
	Serious	0.003	0.468
	Severe	0.0004	0.440
	Critical	0.0002	0.425
	Fatal	0.0001	0.064
	Injured people per crash	0.189	0.340

Pedestrian Crash With Prior Vehicle Maneuver

Typical Scenario: Vehicle is turning left in an urban area, in daylight, under clear weather conditions with a posted speed limit of 35 mph; and encounters a pedestrian in the crosswalk at a signaled intersection.

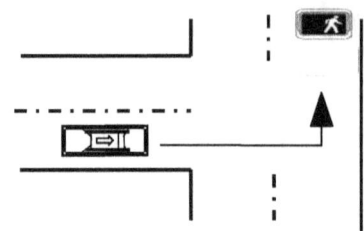

Factor Over-Representation: Urban area, intersection and intersection-related locations, low-speed road, vision obscured, and inattention are over-represented (based on a simple comparison of percentages).

Dynamic Variations: Vehicle is turning right and encounters a pedestrian. The pedestrian is running into the road or playing in the roadway in about 15 percent of overall scenario crashes.

Scenario Severity: Table below quantifies the annual severity of this crash scenario in terms of five different metrics based on 2004 GES statistics. This table also provides the ratios of people involved by maximum injury severity using the KABCO and AIS injury scales. About 2.87 percent of all people involved in this crash scenario suffered high-level MAIS 3+ injuries (serious, severe, critical, or fatal).

	Crash Severity	**Scenario**	**Scenario/All**
	No. of crashes	17,000	0.29%
	No. of vehicles involved	18,000	0.17%
	No. of people involved	41,000	0.27%
Societal Cost	Economic cost	$843,000,000	0.70%
	Functional years lost	24,000	0.88%
KABCO Injury Scale	None	0.545	0.666
	Possible	0.228	2.090
	Non-incapacitating	0.150	3.119
	Incapacitating	0.054	2.799
	Fatal	0.007	4.148
	Unknown	0.016	4.288
	Died prior	-	-
AIS Injury Scale	None	0.558	0.715
	Minor	0.360	1.910
	Moderate	0.053	2.509
	Serious	0.018	2.651
	Severe	0.002	2.877
	Critical	0.001	2.806
	Fatal	0.007	4.111
	Injured people per crash	1.060	1.910

Pedestrian Crash Without Prior Vehicle Maneuver

Typical Scenario: Vehicle is going straight in an urban area, in daylight, under clear weather conditions, with a posted speed limit of 25 mph; and then encounters a pedestrian at a non-junction location.

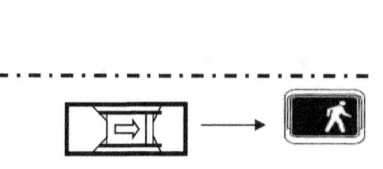

Factor Over-Representation: Dark, adverse weather, non-junction area, low-speed road, vision obscured, and younger driver are over-represented (based on a simple comparison of percentages).

Dynamic Variations: Vehicle is starting in traffic lane or negotiating a curve and encounters a pedestrian. The pedestrian is running into the road in 36 percent of overall scenario crashes. Moreover, the pedestrian is improperly crossing the roadway in 26 percent of overall scenario crashes.

Scenario Severity: Table below quantifies the annual severity of this crash scenario in terms of five different metrics based on 2004 GES statistics. This table also provides the ratios of people involved by maximum injury severity using the KABCO and AIS injury scales. About 5.74 percent of all people involved in this crash scenario suffered high-level MAIS 3+ injuries (serious, severe, critical, or fatal).

	Crash Severity	**Scenario**	**Scenario/All**
	No. of crashes	39,000	0.66%
	No. of vehicles involved	42,000	0.39%
	No. of people involved	98,000	0.65%
Societal Cost	Economic cost	$4,022,000,000	3.36%
	Functional years lost	144,000	5.21%
KABCO Injury Scale	None	0.587	0.717
	Possible	0.124	1.131
	Non-incapacitating	0.131	2.715
	Incapacitating	0.115	5.997
	Fatal	0.025	14.008
	Unknown	0.019	5.236
	Died prior	-	-
AIS Injury Scale	None	0.576	0.738
	Minor	0.305	1.618
	Moderate	0.061	2.899
	Serious	0.026	3.878
	Severe	0.004	4.957
	Critical	0.002	5.462
	Fatal	0.025	13.884
	Injured people per crash	1.055	1.902

Pedalcyclist Crash With Prior Vehicle Maneuver

Typical Scenario: Vehicle is turning right in an urban area, in daylight, under clear weather conditions, with a posted speed limit of 25 mph; and encounters a pedalcyclist at an intersection.

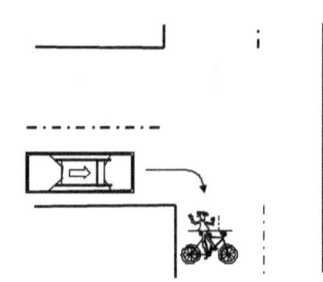

Factor Over-Representation: Clear weather, dry road, intersection and intersection-related locations, low-speed road, vision obscured, inattention, and younger driver are over-represented (based on a simple comparison of percentages).

Dynamic Variations: Vehicle is turning left and encounters a pedalcyclist. The pedalcyclist is in the crosswalk in about one-third of overall scenario crashes. Moreover, the pedalcyclist fails to yield the right-of-way and is riding on the wrong side of the road respectively in about 13 and 24 percent of overall scenario crashes.

Scenario Severity: Table below quantifies the annual severity of this crash scenario in terms of five different metrics based on 2004 GES statistics. This table also provides the ratios of people involved by maximum injury severity using the KABCO and AIS injury scales. About 1.65 percent of all people involved in this crash scenario suffered high-level MAIS 3+ injuries (serious, severe, critical, or fatal).

	Crash Severity	**Scenario**	**Scenario/All**
	No. of crashes	18,000	0.31%
	No. of vehicles involved	19,000	0.18%
	No. of people involved	48,000	0.32%
Societal Cost	Economic cost	$523,000,000	0.44%
	Functional years lost	11,000	0.39%
KABCO Injury Scale	None	0.645	0.788
	Possible	0.126	1.158
	Non-incapacitating	0.189	3.922
	Incapacitating	0.035	1.821
	Fatal	0.0002	0.127
	Unknown	0.005	1.279
	Died prior	-	-
AIS Injury Scale	None	0.631	0.809
	Minor	0.308	1.634
	Moderate	0.044	2.091
	Serious	0.014	2.062
	Severe	0.002	2.039
	Critical	0.001	1.956
	Fatal	0.0002	0.126
	Injured people per crash	0.975	1.757

Pedalcyclist Crash Without Prior Vehicle Maneuver

Typical Scenario: Vehicle is going straight in an urban area, in daylight, under clear weather conditions, with a posted speed limit of 25 mph; and encounters a pedalcyclist at an intersection.

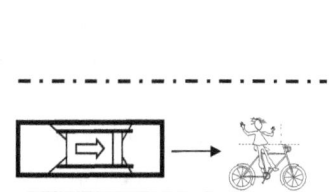

Factor Over-Representation: Clear weather, dry road, intersection, low-speed road, vision obscured, and female driver are over-represented (based on a simple comparison of percentages).

Dynamic Variations: Vehicle is starting in traffic lane and encounters a pedalcyclist. The pedalcyclist fails to yield the right-of-way and is riding on the wrong side of the road respectively in about 46 and 6 percent of overall scenario crashes.

Scenario Severity: Table below quantifies the annual severity of this crash scenario in terms of five different metrics based on 2004 GES statistics. This table also provides the ratios of people involved by maximum injury severity using the KABCO and AIS injury scales. About 3.27 percent of all people involved in this crash scenario suffered high-level MAIS 3+ injuries (serious, severe, critical, or fatal).

	Crash Severity	**Scenario**	**Scenario/All**
	No. of crashes	24,000	0.41%
	No. of vehicles involved	25,000	0.23%
	No. of people involved	58,000	0.39%
Societal Cost	Economic cost	$1,301,000,000	1.09%
	Functional years lost	39,000	1.42%
KABCO Injury Scale	None	0.593	0.726
	Possible	0.134	1.229
	Non-incapacitating	0.184	3.823
	Incapacitating	0.070	3.663
	Fatal	0.009	4.837
	Unknown	0.009	2.518
	Died prior	-	-
AIS Injury Scale	None	0.586	0.751
	Minor	0.327	1.733
	Moderate	0.054	2.585
	Serious	0.020	2.964
	Severe	0.003	3.362
	Critical	0.001	3.537
	Fatal	0.009	4.795
	Injured people per crash	1.003	1.808

Backing Up Into Another Vehicle

Typical Scenario: Vehicle is backing up in an urban area, in daylight, under clear weather conditions, at a driveway or alley location, with a posted speed limit of 25 mph; and collides with another vehicle.

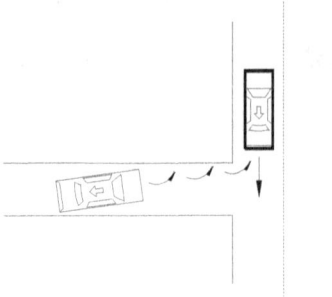

Factor Over-Representation: Daylight, driveway or alley and intersection-related locations, low-speed road, vision obscured, inattention, and younger driver are over-represented (based on a simple comparison of percentages).

Dynamic Variations: Vehicle is leaving a parked position and backs into another vehicle.

Scenario Severity: Table below quantifies the annual severity of this crash scenario in terms of five different metrics based on 2004 GES statistics. This table also provides the ratios of people involved by maximum injury severity using the KABCO and AIS injury scales. About 0.13 percent of all people involved in this crash scenario suffered high-level MAIS 3+ injuries (serious, severe, critical, or fatal).

	Crash Severity	**Scenario**	**Scenario/All**
	No. of crashes	131,000	2.20%
	No. of vehicles involved	261,000	2.44%
	No. of people involved	363,000	2.42%
Societal Cost	Economic cost	$947,000,000	0.79%
	Functional years lost	9,000	0.32%
KABCO Injury Scale	None	0.957	1.170
	Possible	0.034	0.313
	Non-incapacitating	0.007	0.143
	Incapacitating	0.001	0.030
	Fatal	0.00003	0.019
	Unknown	0.001	0.371
	Died prior	-	-
AIS Injury Scale	None	0.892	1.142
	Minor	0.102	0.538
	Moderate	0.006	0.263
	Serious	0.001	0.173
	Severe	0.0001	0.109
	Critical	0.00002	0.058
	Fatal	0.00003	0.019
	Injured people per crash	0.301	0.542

Vehicle(s) Turning – Vehicles Traveling in Same Direction

Typical Scenario: Vehicle is turning left at an intersection in an urban area, in daylight, under clear weather conditions, with a posted speed limit of 35 mph; and then cuts across the path of another vehicle initially traveling in the same direction.

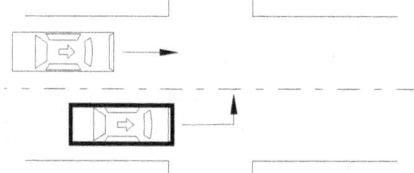

Factor Over-Representation: Clear weather, dry road, low-speed road, and younger driver are over-represented (based on a simple comparison of percentages).

Dynamic Variations: Vehicle is turning right and cuts across the path of another vehicle initially traveling in the same direction.

Scenario Severity: Table below quantifies the annual severity of this crash scenario in terms of five different metrics based on 2004 GES statistics. This table also provides the ratios of people involved by maximum injury severity using the KABCO and AIS injury scales. About 0.44 percent of all people involved in this crash scenario suffered high-level MAIS 3+ injuries (serious, severe, critical, or fatal).

	Crash Severity	**Scenario**	**Scenario/All**
	No. of crashes	222,000	3.73%
	No. of vehicles involved	446,000	4.17%
	No. of people involved	641,000	4.26%
Societal Cost	Economic cost	$2,810,000,000	2.34%
	Functional years lost	47,000	1.68%
KABCO Injury Scale	None	0.900	1.100
	Possible	0.066	0.608
	Non-incapacitating	0.023	0.470
	Incapacitating	0.009	0.455
	Fatal	0.0003	0.190
	Unknown	0.002	0.574
	Died prior	-	-
AIS Injury Scale	None	0.846	1.084
	Minor	0.137	0.728
	Moderate	0.012	0.568
	Serious	0.003	0.521
	Severe	0.0004	0.485
	Critical	0.0002	0.465
	Fatal	0.0003	0.189
	Injured people per crash	0.444	0.801

Vehicle(s) Parking – Vehicles Traveling in Same Direction

Typical Scenario: Vehicle is leaving a parked position in an urban area, in daylight, under clear weather conditions, with a posted speed limit of 25 mph; and encounters another vehicle traveling in the same direction at a non-junction area.

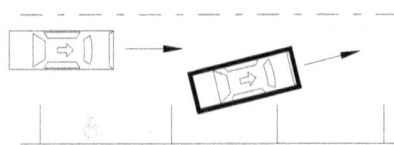

Factor Over-Representation: Adverse weather, non-junction area, low-speed road, inattention, and younger driver are over-represented (based on a simple comparison of percentages).

Dynamic Variations: Vehicle is making a U-turn and encounters a vehicle traveling in the same direction.

Scenario Severity: Table below quantifies the annual severity of this crash scenario in terms of five different metrics based on 2004 GES statistics. This table also provides the ratios of people involved by maximum injury severity using the KABCO and AIS injury scales. About 0.45 percent of all people involved in this crash scenario suffered high-level MAIS 3+ injuries (serious, severe, critical, or fatal).

	Crash Severity	Scenario	Scenario/All
	No. of crashes	48,000	0.81%
	No. of vehicles involved	95,000	0.89%
	No. of people involved	125,000	0.83%
Societal Cost	Economic cost	$623,000,000	0.52%
	Functional years lost	11,000	0.41%
KABCO Injury Scale	None	0.892	1.090
	Possible	0.064	0.582
	Non-incapacitating	0.038	0.781
	Incapacitating	0.004	0.228
	Fatal	0.0009	0.485
	Unknown	0.002	0.543
	Died prior	-	-
AIS Injury Scale	None	0.839	1.074
	Minor	0.144	0.766
	Moderate	0.012	0.588
	Serious	0.003	0.473
	Severe	0.0003	0.376
	Critical	0.0001	0.295
	Fatal	0.0009	0.480
	Injured people per crash	0.426	0.768

Vehicle(s) Changing Lanes – Vehicles Traveling in Same Direction

Typical Scenario: Vehicle is changing lanes in an urban area, in daylight, under clear weather conditions, at a non-junction with a posted speed limit of 55 mph or more; and then encroaches into another vehicle traveling in the same direction.

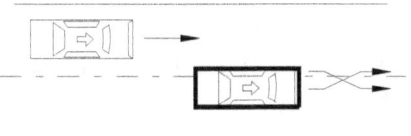

Factor Over-Representation: Non-junction area, high-speed road, inattention, and younger driver are over-represented (based on a simple comparison of percentages).

Dynamic Variations: Vehicle is passing another vehicle and encroaches into another vehicle traveling in the same direction (15% of crashes). Vehicle may also be merging (8% of crashes). When changing lanes or passing, the vehicle is equally as likely to be moving to the right as to the left. On the other hand, the vehicle merges to the left in about 75 percent of the merging crashes.

Scenario Severity: Table below quantifies the annual severity of this crash scenario in terms of five different metrics based on 2004 GES statistics. This table also provides the ratios of people involved by maximum injury severity using the KABCO and AIS injury scales. About 0.42 percent of all people involved in this crash scenario suffered high-level MAIS 3+ injuries (serious, severe, critical, or fatal).

	Crash Severity	Scenario	Scenario/All
	No. of crashes	338,000	5.69%
	No. of vehicles involved	635,000	5.94%
	No. of people involved	884,000	5.88%
Societal Cost	Economic cost	$4,247,000,000	3.54%
	Functional years lost	71,000	2.57%
KABCO Injury Scale	None	0.924	1.129
	Possible	0.048	0.441
	Non-incapacitating	0.017	0.351
	Incapacitating	0.008	0.421
	Fatal	0.0007	0.396
	Unknown	0.002	0.666
	Died prior	-	-
AIS Injury Scale	None	0.864	1.107
	Minor	0.121	0.644
	Moderate	0.010	0.471
	Serious	0.003	0.441
	Severe	0.0004	0.437
	Critical	0.0002	0.419
	Fatal	0.0007	0.392
	Injured people per crash	0.387	0.697

Vehicle(s) Drifting – Vehicles Traveling in Same Direction

Typical Scenario: Vehicle is going straight in an urban area, in daylight, under clear weather conditions, at a non-junction with a posted speed limit of 55 mph or more; and then drifts into an adjacent vehicle traveling in the same direction.

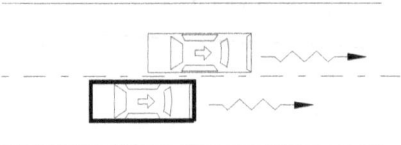

Factor Over-Representation: High-speed road, speeding, and younger driver are over-represented (based on a simple comparison of percentages).

Dynamic Variations: Vehicle drifts into another vehicle stopped in traffic lane.

Scenario Severity: Table below quantifies the annual severity of this crash scenario in terms of five different metrics based on 2004 GES statistics. This table also provides the ratios of people involved by maximum injury severity using the KABCO and AIS injury scales. About 0.58 percent of all people involved in this crash scenario suffered high-level MAIS 3+ injuries (serious, severe, critical, or fatal).

	Crash Severity	**Scenario**	**Scenario/All**
	No. of crashes	98,000	1.65%
	No. of vehicles involved	235,000	2.20%
	No. of people involved	330,000	2.19%
Societal Cost	Economic cost	$1,383,000,000	1.15%
	Functional years lost	37,000	1.32%
KABCO Injury Scale	None	0.893	1.092
	Possible	0.067	0.612
	Non-incapacitating	0.026	0.534
	Incapacitating	0.011	0.598
	Fatal	0.001	0.587
	Unknown	0.001	0.374
	Died prior	-	-
AIS Injury Scale	None	0.841	1.077
	Minor	0.141	0.745
	Moderate	0.013	0.618
	Serious	0.004	0.600
	Severe	0.0005	0.577
	Critical	0.0002	0.590
	Fatal	0.001	0.581
	Injured people per crash	0.413	0.744

Vehicle(s) Making a Maneuver – Vehicles Traveling in Opposite Direction

Typical Scenario: Vehicle is passing another vehicle in a rural area, in daylight, under clear weather conditions, at a non-junction with a posted speed limit of 55 mph or more; and encroaches into another vehicle traveling in the opposite direction.

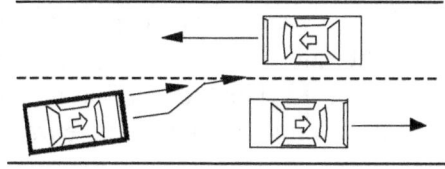

Factor Over-Representation: Dark, adverse weather, rural area, non-junction, high-speed road, alcohol, vision obscured, inattention, speeding, male, and young driver are over-represented (based on a simple comparison of percentages).

Dynamic Variations: Vehicle is changing lanes or in the middle of a corrective maneuver and encroaches into another vehicle traveling in the opposite direction.

Scenario Severity: Table below quantifies the annual severity of this crash scenario in terms of five different metrics based on 2004 GES statistics. This table also provides the ratios of people involved by maximum injury severity using the KABCO and AIS injury scales. About 3.16 percent of all people involved in this crash scenario suffered high-level MAIS 3+ injuries (serious, severe, critical, or fatal).

	Crash Severity	**Scenario**	**Scenario/All**
	No. of crashes	15,000	0.26%
	No. of vehicles involved	30,000	0.28%
	No. of people involved	40,000	0.27%
Societal Cost	Economic cost	$943,000,000	0.79%
	Functional years lost	32,000	1.14%
KABCO Injury Scale	None	0.710	0.868
	Possible	0.130	1.189
	Non-incapacitating	0.079	1.649
	Incapacitating	0.063	3.305
	Fatal	0.013	7.125
	Unknown	0.005	1.251
	Died prior	-	-
AIS Injury Scale	None	0.687	0.881
	Minor	0.243	1.286
	Moderate	0.039	1.833
	Serious	0.015	2.288
	Severe	0.0022	2.684
	Critical	0.0012	3.031
	Fatal	0.013	7.062
	Injured people per crash	0.816	1.470

Vehicle(s) Not Making a Maneuver – Vehicles Traveling in Opposite Direction

Typical Scenario: Vehicle is going straight in a rural area, in daylight, under clear weather conditions, at a non-junction with a posted speed limit of 55 mph or more; and drifts and encroaches into another vehicle traveling in the opposite direction.

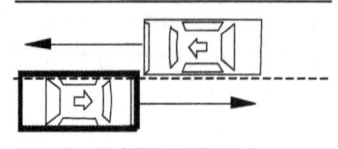

Factor Over-Representation: Dark, adverse weather, wet or slippery road surface, non-level road, rural area, non-junction, alcohol, male, and younger driver are over-represented (based on a simple comparison of percentages).

Dynamic Variations: Vehicle is negotiating a curve and then drifts and encroaches into another vehicle traveling in the opposite direction. About 42 percent of overall scenario crashes occur on curves.

Scenario Severity: Table below quantifies the annual severity of this crash scenario in terms of five different metrics based on 2004 GES statistics. This table also provides the ratios of people involved by maximum injury severity using the KABCO and AIS injury scales. About 2.58 percent of all people involved in this crash scenario suffered high-level MAIS 3+ injuries (serious, severe, critical, or fatal).

	Crash Severity	Scenario	Scenario/All
	No. of crashes	124,000	2.08%
	No. of vehicles involved	232,000	2.17%
	No. of people involved	330,000	2.20%
Societal Cost	Economic cost	$6,407,000,000	5.35%
	Functional years lost	206,000	7.44%
KABCO Injury Scale	None	0.723	0.884
	Possible	0.119	1.086
	Non-incapacitating	0.092	1.906
	Incapacitating	0.049	2.536
	Fatal	0.010	5.448
	Unknown	0.008	2.122
	Died prior	-	-
AIS Injury Scale	None	0.698	0.894
	Minor	0.240	1.274
	Moderate	0.036	1.701
	Serious	0.013	1.972
	Severe	0.002	2.265
	Critical	0.0009	2.403
	Fatal	0.010	5.400
	Injured people per crash	0.806	1.452

Following Vehicle Making a Maneuver and Approaching Lead Vehicle

Typical Scenario: Vehicle is changing lanes or passing in an urban area, in daylight, under clear weather conditions, at a non-junction with a posted speed limit of 55 mph; and closes in on a lead vehicle.

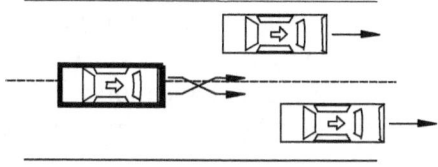

Factor Over-Representation: Intersection-related location, inattention, speeding, and younger driver are over-represented (based on a simple comparison of percentages).

Dynamic Variations: Vehicle is turning right and then closes in on a lead vehicle (22% of crashes).

Scenario Severity: Table below quantifies the annual severity of this crash scenario in terms of five different metrics based on 2004 GES statistics. This table also provides the ratios of people involved by maximum injury severity using the KABCO and AIS injury scales. About 0.50 percent of all people involved in this crash scenario suffered high-level MAIS 3+ injuries (serious, severe, critical, or fatal).

	Crash Severity	**Scenario**	**Scenario/All**
	No. of crashes	85,000	1.44%
	No. of vehicles involved	180,000	1.69%
	No. of people involved	249,000	1.66%
Societal Cost	Economic cost	$1,212,000,000	1.01%
	Functional years lost	18,000	0.67%
KABCO Injury Scale	None	0.860	1.052
	Possible	0.103	0.946
	Non-incapacitating	0.023	0.482
	Incapacitating	0.009	0.487
	Fatal	0.0001	0.053
	Unknown	0.004	1.049
	Died prior	-	-
AIS Injury Scale	None	0.817	1.047
	Minor	0.163	0.864
	Moderate	0.015	0.707
	Serious	0.004	0.632
	Severe	0.0005	0.573
	Critical	0.0002	0.516
	Fatal	0.0001	0.053
	Injured people per crash	0.533	0.962

Following Vehicle Approaching an Accelerating Lead Vehicle

Typical Scenario: Vehicle is going straight in an urban area, in daylight, under clear weather conditions, at an intersection-related location with a posted speed limit of 45 mph; and closes in on an accelerating lead vehicle.

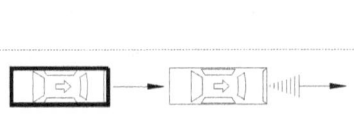

Factor Over-Representation: Dry road, intersection-related, high-speed road, traffic signal, inattention, speeding, female, and younger driver are over-represented (based on a simple comparison of percentages).

Dynamic Variations: Vehicle is starting in traffic lane and then closes in on an accelerating lead vehicle (34% of crashes).

Scenario Severity: Table below quantifies the annual severity of this crash scenario in terms of five different metrics based on 2004 GES statistics. This table also provides the ratios of people involved by maximum injury severity using the KABCO and AIS injury scales. About 0.55 percent of all people involved in this crash scenario suffered high-level MAIS 3+ injuries (serious, severe, critical, or fatal).

	Crash Severity	Scenario	Scenario/All
	No. of crashes	19,000	0.32%
	No. of vehicles involved	40,000	0.38%
	No. of people involved	54,000	0.36%
Societal Cost	Economic cost	$273,000,000	0.23%
	Functional years lost	4,000	0.15%
KABCO Injury Scale	None	0.865	1.058
	Possible	0.088	0.802
	Non-incapacitating	0.035	0.724
	Incapacitating	0.012	0.625
	Fatal	0.0001	0.057
	Unknown	0.000	0.000
	Died prior	-	-
AIS Injury Scale	None	0.819	1.049
	Minor	0.160	0.848
	Moderate	0.015	0.733
	Serious	0.005	0.690
	Severe	0.0005	0.611
	Critical	0.0003	0.633
	Fatal	0.0001	0.056
	Injured people per crash	0.518	0.934

Following Vehicle Approaching Lead Vehicle Moving at Lower Constant Speed

Typical Scenario: Vehicle is going straight in an urban area, in daylight, under clear weather conditions, at a non-junction with a posted speed limit of 55 mph or more; and closes in on a lead vehicle moving at lower constant speed.

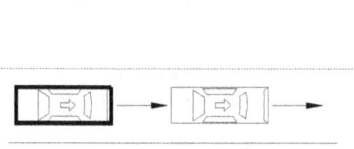

Factor Over-Representation: Non-junction location, high-speed road, inattention, speeding, and younger driver are over-represented (based on a simple comparison of percentages).

Dynamic Variations: Vehicle is decelerating in traffic lane and then closes in on a lead vehicle moving at lower constant speed.

Scenario Severity: Table below quantifies the annual severity of this crash scenario in terms of five different metrics based on 2004 GES statistics. This table also provides the ratios of people involved by maximum injury severity using the KABCO and AIS injury scales. About 0.71 percent of all people involved in this crash scenario suffered high-level MAIS 3+ injuries (serious, severe, critical, or fatal).

	Crash Severity	**Scenario**	**Scenario/All**
	No. of crashes	210,000	3.53%
	No. of vehicles involved	445,000	4.16%
	No. of people involved	612,000	4.07%
Societal Cost	Economic cost	$3,910,000,000	3.26%
	Functional years lost	78,000	2.81%
KABCO Injury Scale	None	0.836	1.022
	Possible	0.116	1.065
	Non-incapacitating	0.031	0.652
	Incapacitating	0.013	0.694
	Fatal	0.001	0.548
	Unknown	0.002	0.593
	Died prior	-	-
AIS Injury Scale	None	0.797	1.022
	Minor	0.178	0.943
	Moderate	0.018	0.836
	Serious	0.005	0.785
	Severe	0.0006	0.714
	Critical	0.0003	0.707
	Fatal	0.001	0.543
	Injured people per crash	0.592	1.066

Following Vehicle Approaching a Decelerating Lead Vehicle

Typical Scenario: Vehicle is going straight and following another lead vehicle in a rural area, in daylight, under clear weather conditions, at a non-junction with a posted speed limit of 55 mph or more; and the lead vehicle suddenly decelerates.

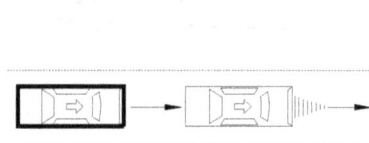

Factor Over-Representation: Daylight, adverse weather, rural area, intersection-related, high-speed road, inattention, speeding, and younger driver are over-represented (based on a simple comparison of percentages).

Dynamic Variations: Vehicle is decelerating in traffic lane and then closes in on a decelerating lead vehicle (11% of crashes).

Scenario Severity: Table below quantifies the annual severity of this crash scenario in terms of five different metrics based on 2004 GES statistics. This table also provides the ratios of people involved by maximum injury severity using the KABCO and AIS injury scales. About 0.49 percent of all people involved in this crash scenario suffered high-level MAIS 3+ injuries (serious, severe, critical, or fatal).

	Crash Severity	**Scenario**	**Scenario/All**
	No. of crashes	428,000	7.20%
	No. of vehicles involved	936,000	8.76%
	No. of people involved	1,283,000	8.54%
Societal Cost	Economic cost	$6,390,000,000	5.33%
	Functional years lost	100,000	3.62%
KABCO Injury Scale	None	0.856	1.047
	Possible	0.112	1.026
	Non-incapacitating	0.022	0.455
	Incapacitating	0.009	0.452
	Fatal	0.0003	0.140
	Unknown	0.001	0.293
	Died prior	-	-
AIS Injury Scale	None	0.815	1.044
	Minor	0.166	0.878
	Moderate	0.015	0.698
	Serious	0.004	0.611
	Severe	0.0004	0.495
	Critical	0.0002	0.479
	Fatal	0.0003	0.139
	Injured people per crash	0.555	1.001

Following Vehicle Approaching a Stopped Lead Vehicle

Typical Scenario: Vehicle is going straight in an urban area, in daylight, under clear weather conditions, at an intersection-related location with a posted speed limit of 35 mph; and closes in on a stopped lead vehicle.

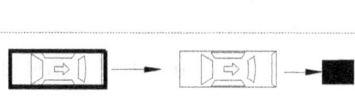

Factor Over-Representation: Rural area, intersection-related, inattention, speeding and younger driver are over-represented (based on a simple comparison of percentages).

Dynamic Variations: Vehicle is decelerating in traffic lane and closes in on a stopped lead vehicle (12% of crashes). Vehicle may also be starting in traffic lane and closes in on a stopped lead vehicle (8% of crashes). In about 50 percent of the lead-vehicle-stopped crashes, the lead vehicle first decelerates to a stop and is struck afterwards by the following vehicle. This typically happens in the presence of a traffic control device or the lead vehicle is slowing down to make a turn. Thus, this particular scenario overlaps with the lead vehicle-decelerating scenario.

Scenario Severity: Table below quantifies the annual severity of this crash scenario in terms of five different metrics based on 2004 GES statistics. This table also provides the ratios of people involved by maximum injury severity using the KABCO and AIS injury scales. About 0.50 percent of all people involved in this crash scenario suffered high-level MAIS 3+ injuries (serious, severe, critical, or fatal).

	Crash Severity	**Scenario**	**Scenario/All**
	No. of crashes	975,000	16.41%
	No. of vehicles involved	2,162,000	20.21%
	No. of people involved	3,032,000	20.18%
Societal Cost	Economic cost	$15,388,000,000	12.84%
	Functional years lost	240,000	8.69%
KABCO Injury Scale	None	0.844	1.032
	Possible	0.121	1.108
	Non-incapacitating	0.023	0.482
	Incapacitating	0.008	0.397
	Fatal	0.0002	0.128
	Unknown	0.004	0.995
	Died prior	0.00005	1.921
AIS Injury Scale	None	0.806	1.032
	Minor	0.174	0.920
	Moderate	0.016	0.738
	Serious	0.004	0.627
	Severe	0.0004	0.522
	Critical	0.0002	0.446
	Fatal	0.0002	0.127
	Injured people per crash	0.604	1.088

Left Turn across Path from Opposite Directions at Signalized Junctions

Typical Scenario: Vehicle is turning left in an urban area, in daylight, under clear weather conditions, at a signalized intersection with a posted speed limit of 35 mph; and cuts across the path of another vehicle straight crossing from an opposite direction.

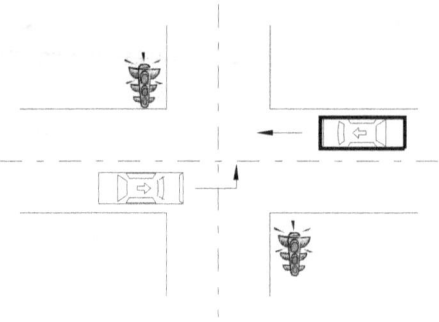

Factor Over-Representation: Intersection, low-speed road, vision obscured, inattention, female, and younger driver are over-represented (based on a simple comparison of percentages).

Dynamic Variations: Vehicle is turning left across the path of another vehicle that is also turning left from the opposite direction.

Scenario Severity: Table below quantifies the annual severity of this crash scenario in terms of five different metrics based on 2004 GES statistics. This table also provides the ratios of people involved by maximum injury severity using the KABCO and AIS injury scales. About 1.16 percent of all people involved in this crash scenario suffered high-level MAIS 3+ injuries (serious, severe, critical, or fatal).

	Crash Severity	**Scenario**	**Scenario/All**
	No. of crashes	220,000	3.71%
	No. of vehicles involved	457,000	4.28%
	No. of people involved	664,000	4.42%
Societal Cost	Economic cost	$5,749,000,000	4.80%
	Functional years lost	121,000	4.36%
KABCO Injury Scale	None	0.753	0.920
	Possible	0.144	1.314
	Non-incapacitating	0.074	1.526
	Incapacitating	0.025	1.319
	Fatal	0.001	0.531
	Unknown	0.004	1.043
	Died prior	-	-
AIS Injury Scale	None	0.729	0.934
	Minor	0.232	1.228
	Moderate	0.028	1.336
	Serious	0.009	1.341
	Severe	0.001	1.325
	Critical	0.0005	1.331
	Fatal	0.001	0.527
	Injured people per crash	0.818	1.474

Vehicle Turning Right at Signalized Junctions

Typical Scenario: Vehicle is turning right in an urban area, in daylight, under clear weather conditions, at a signalized intersection with a posted speed limit of 35 mph; and turns into the same direction of another vehicle crossing straight initially from a lateral direction.

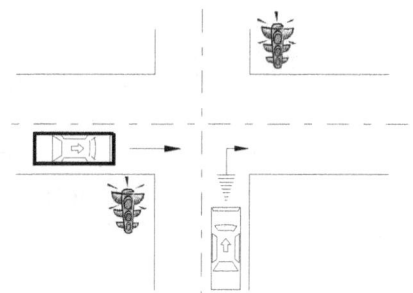

Factor Over-Representation: Adverse weather, intersection or intersection-related locations, low-speed road, vision obscured, and younger and older drivers are over-represented (based on a simple comparison of percentages).

Dynamic Variations: Vehicle is turning right at a signalized intersection and then turns into the opposite direction of another vehicle traveling or stopped initially from a lateral direction.

Scenario Severity: Table below quantifies the annual severity of this crash scenario in terms of five different metrics based on 2004 GES statistics. This table also provides the ratios of people involved by maximum injury severity using the KABCO and AIS injury scales. About 0.27 percent of all people involved in this crash scenario suffered high-level MAIS 3+ injuries (serious, severe, critical, or fatal).

	Crash Severity	**Scenario**	**Scenario/All**
	No. of crashes	35,000	0.59%
	No. of vehicles involved	71,000	0.66%
	No. of people involved	98,000	0.65%
Societal Cost	Economic cost	$355,000,000	0.30%
	Functional years lost	4,000	0.15%
KABCO Injury Scale	None	0.900	1.100
	Possible	0.076	0.698
	Non-incapacitating	0.019	0.400
	Incapacitating	0.002	0.108
	Fatal	-	-
	Unknown	0.002	0.617
	Died prior	-	-
AIS Injury Scale	None	0.848	1.087
	Minor	0.139	0.735
	Moderate	0.010	0.493
	Serious	0.002	0.364
	Severe	0.0002	0.251
	Critical	0.0001	0.168
	Fatal	-	-
	Injured people per crash	0.425	0.767

Left Turn Across Path From Opposite Directions at Non-Signalized Junctions

Typical Scenario: Vehicle is turning left, in daylight, under clear weather conditions, at an intersection without traffic controls, with a posted speed limit of 35 mph; and then cuts across the path of another vehicle traveling from the opposite direction.

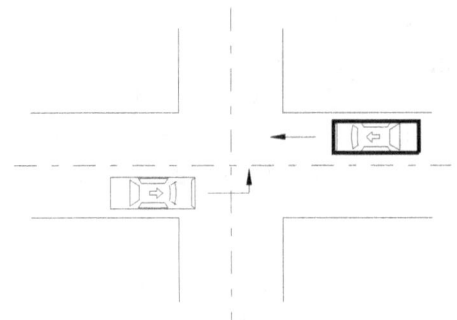

Factor Over-Representation: Rural area, intersection and driveway/alley locations, low-speed road, vision obscured, inattention, and younger and older drivers are over-represented (based on a simple comparison of percentages).

Dynamic Variations: Two vehicles are traveling in opposite directions and then both vehicles may turn left across their paths.

Scenario Severity: Table below quantifies the annual severity of this crash scenario in terms of five different metrics based on 2004 GES statistics. This table also provides the ratios of people involved by maximum injury severity using the KABCO and AIS injury scales. About 1.24 percent of all people involved in this crash scenario suffered high-level MAIS 3+ injuries (serious, severe, critical, or fatal).

	Crash Severity	Scenario	Scenario/All
	No. of crashes	190,000	3.19%
	No. of vehicles involved	389,000	3.64%
	No. of people involved	558,000	3.71%
Societal Cost	Economic cost	$5,137,000,000	4.29%
	Functional years lost	113,000	4.09%
KABCO Injury Scale	None	0.749	0.916
	Possible	0.144	1.322
	Non-incapacitating	0.073	1.522
	Incapacitating	0.027	1.412
	Fatal	0.001	0.737
	Unknown	0.005	1.275
	Died prior	-	-
AIS Injury Scale	None	0.726	0.930
	Minor	0.233	1.237
	Moderate	0.029	1.368
	Serious	0.009	1.392
	Severe	0.001	1.405
	Critical	0.0006	1.414
	Fatal	0.001	0.731
	Injured people per crash	0.806	1.453

Straight Crossing Paths at Non-Signalized Junctions

Typical Scenario: Vehicle stops at a stop sign in an urban area, in daylight, under clear weather conditions, at an intersection with a posted speed limit of 25 mph; and then proceeds against lateral crossing traffic.

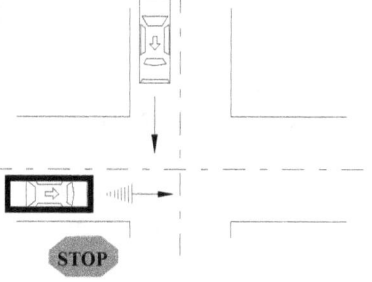

Factor Over-Representation: Rural area, low-speed road, vision obscured, female, and younger and older drivers are over-represented (based on a simple comparison of percentages).

Dynamic Variations: Vehicle is going straight through an uncontrolled intersection and then cuts across the path of another straight crossing vehicle from lateral direction. Another scenario involves both vehicles first stopping and then proceeding on straight crossing paths.

Scenario Severity: Table below quantifies the annual severity of this crash scenario in terms of five different metrics based on 2004 GES statistics. This table also provides the ratios of people involved by maximum injury severity using the KABCO and AIS injury scales. About 1.21 percent of all people involved in this crash scenario suffered high-level MAIS 3+ injuries (serious, severe, critical, or fatal).

	Crash Severity	**Scenario**	**Scenario/All**
	No. of crashes	264,000	4.44%
	No. of vehicles involved	535,000	5.00%
	No. of people involved	765,000	5.09%
Societal Cost	Economic cost	$7,290,000,000	6.08%
	Functional years lost	174,000	6.29%
KABCO Injury Scale	None	0.769	0.940
	Possible	0.139	1.276
	Non-incapacitating	0.062	1.279
	Incapacitating	0.024	1.245
	Fatal	0.002	1.252
	Unknown	0.004	1.103
	Died prior	-	-
AIS Injury Scale	None	0.742	0.951
	Minor	0.220	1.166
	Moderate	0.026	1.237
	Serious	0.008	1.245
	Severe	0.001	1.238
	Critical	0.0005	1.246
	Fatal	0.002	1.241
	Injured people per crash	0.748	1.348

Vehicle(s) Turning at Non-Signalized Junctions

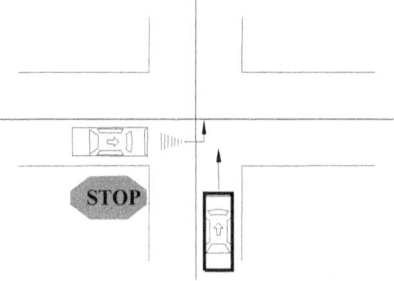

Typical Scenario: Vehicle stops at a stop sign in a rural area, in daylight, under clear weather conditions, at an intersection with a posted speed limit of 35 mph; and proceeds to turn left against lateral crossing traffic.

Factor Over-Representation: Rural area, intersection and driveway/alley locations, low-speed road, vision obscured, inattention, female, and younger and older drivers are over-represented (based on a simple comparison of percentages).

Dynamic Variations: Vehicle stops at a stop sign and then proceeds to turn right against lateral crossing traffic.

Scenario Severity: Table below quantifies the annual severity of this crash scenario in terms of five different metrics based on 2004 GES statistics. This table also provides the ratios of people involved by maximum injury severity using the KABCO and AIS injury scales. About 0.71 percent of all people involved in this crash scenario suffered high-level MAIS 3+ injuries (serious, severe, critical, or fatal).

	Crash Severity	**Scenario**	**Scenario/All**
	No. of crashes	435,000	7.32%
	No. of vehicles involved	872,000	8.15%
	No. of people involved	1,212,000	8.07%
Societal Cost	Economic cost	$7,343,000,000	6.13%
	Functional years lost	138,000	5.00%
KABCO Injury Scale	None	0.843	1.030
	Possible	0.101	0.925
	Non-incapacitating	0.038	0.788
	Incapacitating	0.015	0.778
	Fatal	0.001	0.331
	Unknown	0.003	0.736
	Died prior	-	-
AIS Injury Scale	None	0.801	1.027
	Minor	0.174	0.921
	Moderate	0.018	0.851
	Serious	0.006	0.823
	Severe	0.001	0.790
	Critical	0.0003	0.784
	Fatal	0.001	0.328
	Injured people per crash	0.554	0.998

Vehicle Taking Evasive Action With Prior Vehicle Maneuver

Typical Scenario: Vehicle is turning left at an intersection-related location, in an urban area, in daylight, under clear weather conditions, with a posted speed limit of 35 mph; and takes an evasive action to avoid an obstacle.

Factor Over-Representation: Dark, urban area, intersection-related location, and younger driver are over-represented (based on a simple comparison of percentages).

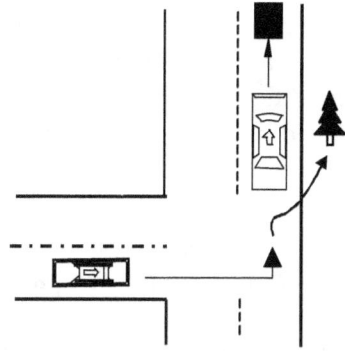

Dynamic Variations: Vehicle is passing, turning right, or changing lanes and then takes an evasive action to avoid an obstacle. The first harmful event occurs on the road in 66 percent of overall scenario crashes and off the road or shoulder/parking lane in 32 percent of the crashes.

Scenario Severity: Table below quantifies the annual severity of this crash scenario in terms of five different metrics based on 2004 GES statistics. This table also provides the ratios of people involved by maximum injury severity using the KABCO and AIS injury scales. About 0.64 percent of all people involved in this crash scenario suffered high-level MAIS 3+ injuries (serious, severe, critical, or fatal).

	Crash Severity	Scenario	Scenario/All
	No. of crashes	13,000	0.22%
	No. of vehicles involved	25,000	0.23%
	No. of people involved	36,000	0.24%
Societal Cost	Economic cost	$198,000,000	0.17%
	Functional years lost	4,000	0.13%
KABCO Injury Scale	None	0.864	1.057
	Possible	0.098	0.895
	Non-incapacitating	0.022	0.452
	Incapacitating	0.016	0.812
	Fatal	0.001	0.293
	Unknown	-	-
	Died prior	-	-
AIS Injury Scale	None	0.820	1.050
	Minor	0.158	0.840
	Moderate	0.015	0.735
	Serious	0.005	0.743
	Severe	0.001	0.703
	Critical	0.0003	0.775
	Fatal	0.001	0.290
	Injured people per crash	0.496	0.895

Vehicle Taking Evasive Action Without Prior Vehicle Maneuver

Typical Scenario: Vehicle is going straight in an urban area, in daylight, under clear weather conditions, at a non-junction location with a posted speed limit of 35 mph; and takes an evasive action to avoid an obstacle.

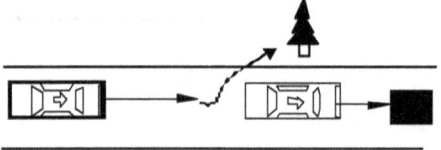

Factor Over-Representation: Driveway/alley and younger driver are over-represented (based on a simple comparison of percentages).

Dynamic Variations: The first harmful event occurs on the road in 65 percent of overall scenario crashes and off the road or shoulder/parking lane in 34 percent of the crashes.

Scenario Severity: Table below quantifies the annual severity of this crash scenario in terms of five different metrics based on 2004 GES statistics. This table also provides the ratios of people involved by maximum injury severity using the KABCO and AIS injury scales. About 1.23 percent of all people involved in this crash scenario suffered high-level MAIS 3+ injuries (serious, severe, critical, or fatal).

	Crash Severity	Scenario	Scenario/All
	No. of crashes	56,000	0.95%
	No. of vehicles involved	99,000	0.93%
	No. of people involved	137,000	0.91%
Societal Cost	Economic cost	$1,349,000,000	1.13%
	Functional years lost	36,000	1.31%
KABCO Injury Scale	None	0.824	1.007
	Possible	0.086	0.789
	Non-incapacitating	0.058	1.201
	Incapacitating	0.023	1.217
	Fatal	0.003	1.917
	Unknown	0.005	1.438
	Died prior	-	-
AIS Injury Scale	None	0.782	1.002
	Minor	0.183	0.972
	Moderate	0.022	1.051
	Serious	0.007	1.105
	Severe	0.001	1.192
	Critical	0.0005	1.196
	Fatal	0.003	1.900
	Injured people per crash	0.530	0.956

Non-Collision Incident

Typical Scenario: Vehicle is going straight in a rural area, in daylight, under clear weather conditions, at a non-junction location with a posted speed limit of over 55 mph; and then fire starts.

Factor Over-Representation: Clear weather, dry road, rural area, non-junction, high-speed road, and vehicle contributing factors are over-represented (based on a simple comparison of percentages).

Dynamic Variations: Vehicle is negotiating a curve and has a non-collision incident. The first harmful event occurs on the road in 90 percent of overall scenario crashes and off the road or shoulder/parking lane in ten percent of the crashes. In this overall scenario, the first harmful events cited are fire or explosion (26%), pavement surface irregularities such as potholes (13%), injured in vehicle or fell from vehicle (10%), thrown or falling object (7%), and other non-collision events. Moreover, this scenario experiences many vehicle-contributing factors such as trailer hitch (10%), tires (9%), power train (7%), wheels (6%), brakes (2%), body or doors (2%), and exhaust system (1%).

Scenario Severity: Table below quantifies the annual severity of this crash scenario in terms of five different metrics based on 2004 GES statistics. This table also provides the ratios of people involved by maximum injury severity using the KABCO and AIS injury scales. About 0.56 percent of all people involved in this crash scenario suffered high-level MAIS 3+ injuries (serious, severe, critical, or fatal).

	Crash Severity	**Scenario**	**Scenario/All**
	No. of crashes	46,000	0.77%
	No. of vehicles involved	82,000	0.77%
	No. of people involved	112,000	0.75%
Societal Cost	Economic cost	$592,000,000	0.49%
	Functional years lost	13,000	0.45%
KABCO Injury Scale	None	0.920	1.125
	Possible	0.038	0.350
	Non-incapacitating	0.028	0.576
	Incapacitating	0.012	0.622
	Fatal	0.001	0.666
	Unknown	0.001	0.148
	Died prior	-	-
AIS Injury Scale	None	0.860	1.101
	Minor	0.123	0.653
	Moderate	0.011	0.540
	Serious	0.004	0.551
	Severe	0.000	0.560
	Critical	0.0002	0.598
	Fatal	0.001	0.660
	Injured people per crash	0.342	0.617

Vehicle Contacting Object with Prior Vehicle Maneuver

Typical Scenario: Vehicle is leaving a parked position at night, in an urban area, under clear weather conditions, at a non-junction location with a posted speed limit of 25 mph; and collides with an object on road shoulder or parking lane.

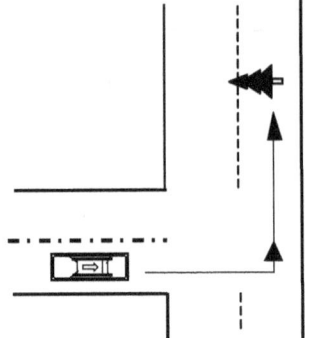

Factor Over-Representation: Dark, wet/slippery road, urban area, non-junction, low-speed road, alcohol, younger driver (71%), and hit-and-run are over-represented (based on a simple comparison of percentages).

Dynamic Variations: Vehicle is turning right and collides with an object. The first harmful event occurs on the road shoulder or parking lane in 64 percent of overall scenario crashes and off the road in 30 percent of the crashes. The first harmful events that are commonly cited are parked motor vehicle (67%) and post, pole, or support (10%).

Scenario Severity: Table below quantifies the annual severity of this crash scenario in terms of five different metrics based on 2004 GES statistics. This table also provides the ratios of people involved by maximum injury severity using the KABCO and AIS injury scales. About 0.35 percent of all people involved in this crash scenario suffered high-level MAIS 3+ injuries (serious, severe, critical, or fatal).

	Crash Severity	Scenario	Scenario/All
	No. of crashes	30,000	0.51%
	No. of vehicles involved	30,000	0.28%
	No. of people involved	34,000	0.23%
Societal Cost	Economic cost	$155,000,000	0.13%
	Functional years lost	3,000	0.10%
KABCO Injury Scale	None	0.957	1.170
	Possible	0.022	0.201
	Non-incapacitating	0.013	0.270
	Incapacitating	0.005	0.280
	Fatal	0.001	0.641
	Unknown	0.002	0.457
	Died prior	-	-
AIS Injury Scale	None	0.890	1.140
	Minor	0.100	0.531
	Moderate	0.007	0.325
	Serious	0.002	0.294
	Severe	0.0002	0.293
	Critical	0.0001	0.277
	Fatal	0.001	0.636
	Injured people per crash	0.125	0.226

Vehicle Contacting Object Without Prior Vehicle Maneuver

Typical Scenario: Vehicle is going straight in a rural area, at night, under clear weather conditions, at a non-junction location with a posted speed limit of 55 mph or more; and collides with an object on the road.

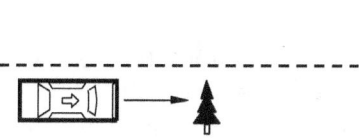

Factor Over-Representation: Dark, rural area, non-junction, high-speed road, alcohol, younger driver, rollover, and hit-and-run are over-represented (based on a simple comparison of percentages).

Dynamic Variations: Vehicle is negotiating a curve and collides with an object. The first harmful event occurs on the road in 54 percent of overall scenario crashes, and on shoulder/parking lane and off the road respectively in 14 and 30 percent of the crashes. The first harmful events that are commonly cited are parked motor vehicle (15%), post, pole, or support (8%), tree (6%), and culvert or ditch (4%). Many objects were coded as "other".

Scenario Severity: Table below quantifies the annual severity of this crash scenario in terms of five different metrics based on 2004 GES statistics. This table also provides the ratios of people involved by maximum injury severity using the KABCO and AIS injury scales. About 1.12 percent of all people involved in this crash scenario suffered high-level MAIS 3+ injuries (serious, severe, critical, or fatal).

	Crash Severity	Scenario	Scenario/All
	No. of crashes	55,000	0.92%
	No. of vehicles involved	55,000	0.51%
	No. of people involved	76,000	0.51%
Societal Cost	Economic cost	$687,000,000	0.57%
	Functional years lost	19,000	0.68%
KABCO Injury Scale	None	0.861	1.052
	Possible	0.069	0.629
	Non-incapacitating	0.042	0.875
	Incapacitating	0.024	1.243
	Fatal	0.003	1.839
	Unknown	0.001	0.319
	Died prior	-	-
AIS Injury Scale	None	0.812	1.040
	Minor	0.158	0.840
	Moderate	0.019	0.881
	Serious	0.007	0.983
	Severe	0.0009	1.061
	Critical	0.0005	1.169
	Fatal	0.003	1.823
	Injured people per crash	0.263	0.474

Other

Other scenarios include on-road rollover, no driver present, hit-and-run, and crash types without any details or specifics. These crashes mostly occur in daylight, under clear weather conditions, dry road surface, straight road, in an urban area, at a non-junction location with a posted speed limit of 25 mph. Vehicle is going straight and encounters a critical event. First harmful event happens on the road.

Factor Over-Representation: Dark, driveway or alley location, low-speed road, rollover, no driver present, hit-and-run, and making a U-turn are over-represented (based on a simple comparison of percentages).

Scenario Severity: Table below quantifies the annual severity of this crash scenario in terms of five different metrics based on 2004 GES statistics. This table also provides the ratios of people involved by maximum injury severity using the KABCO and AIS injury scales. About 1.16 percent of all people involved in this crash scenario suffered high-level MAIS 3+ injuries (serious, severe, critical, or fatal).

	Crash Severity	**Scenario**	**Scenario/All**
	No. of crashes	36,000	0.60%
	No. of vehicles involved	65,000	0.61%
	No. of people involved	78,000	0.52%
Societal Cost	Economic cost	$764,000,000	0.64%
	Functional years lost	21,000	0.75%
KABCO Injury Scale	None	0.855	1.045
	Possible	0.073	0.670
	Non-incapacitating	0.042	0.865
	Incapacitating	0.022	1.143
	Fatal	0.004	2.105
	Unknown	0.005	1.253
	Died prior	-	-
AIS Injury Scale	None	0.807	1.034
	Minor	0.162	0.861
	Moderate	0.019	0.892
	Serious	0.006	0.966
	Severe	0.0009	1.067
	Critical	0.0004	1.098
	Fatal	0.004	2.105
	Injured people per crash	0.418	0.754

5. MAPPING TO NEW PRE-CRASH SCENARIO TYPOLOGY

5.1. Mapping of a Sample of Police-Reported Crashes

A sample of 236 crash police reports was obtained from the department of motor vehicles in the State of Massachusetts. The dates of these reports spanned from September 2004 through March 2005. It should be noted that this time period in Massachusetts covers the severe winter months (November – March), which experienced a substantial amount of snowfall. Each of these police reports was carefully reviewed and assigned to each of the pre-crash scenarios of the new typology. All of them were successfully mapped to this new pre-crash scenario typology as shown in Table 16, except for one crash (other) in which a car being towed by a truck sideswiped six parallel-parked cars. Six scenarios were represented by at least 10 cases, which are listed below by a descending order of number of cases:

1. Lead vehicle stopped: 40 cases (17%)
2. Control loss without prior vehicle action: 21 cases (9%)
3. Control loss with prior vehicle action: 16 cases (7%)
4. Lead vehicle decelerating: 13 cases (6%)
5. Vehicle(s) turning at non-signalized intersections: 10 cases (4%)
6. Backing up into another vehicle: 10 cases (4%)

It is interesting to note that the first two scenarios listed above actually correspond to the top two most-frequent scenarios in the United States as indicated in Table 13. Moreover, the "lead vehicle decelerating" scenario and the "vehicle(s) turning at non-signalized intersections" scenario in the list shown above are ranked respectively fourth and third in the United States.

Table 16. Mapping of a Sample of Crash Reports to New Pre-Crash Scenario

No.	Pre-Crash Scenario	No. Cases	Pct. Cases
1	Vehicle Failure	3	1.3%
2	Control Loss With Prior Vehicle Action	16	6.8%
3	Control Loss Without Prior Vehicle Action	21	8.9%
4	Running Red Light	8	3.4%
5	Running Stop Sign	7	3.0%
6	Road Edge Departure With Prior Vehicle Maneuver	2	0.8%
7	Road Edge Departure Without Prior Vehicle Maneuver	5	2.1%
8	Road Edge Departure While Backing Up	2	0.8%
9	Animal Crash With Prior Vehicle Maneuver	1	0.4%
10	Animal Crash Without Prior Vehicle Maneuver	4	1.7%
11	Pedestrian Crash With Prior Vehicle Maneuver	3	1.3%
12	Pedestrian Crash Without Prior Vehicle Maneuver	1	0.4%
13	Pedalcyclist Crash With Prior Vehicle Maneuver	0	0.0%
14	Pedalcyclist Crash Without Prior Vehicle Maneuver	0	0.0%
15	Backing Up Into Another Vehicle	10	4.2%
16	Vehicle(s) Turning – Same Direction	6	2.5%
17	Vehicle(s) Parking – Same Direction	3	1.3%
18	Vehicle(s) Changing Lanes – Same Direction	9	3.8%
19	Vehicle(s) Drifting – Same Direction	8	3.4%
20	Vehicle(s) Making a Maneuver – Opposite Direction	2	0.8%
21	Vehicle(s) Not Making a Maneuver – Opposite Direction	5	2.1%
22	Following Vehicle Making a Maneuver	1	0.4%
23	Lead Vehicle Accelerating	1	0.4%
24	Lead Vehicle Moving at Lower Constant Speed	5	2.1%
25	Lead Vehicle Decelerating	13	5.5%
26	Lead Vehicle Stopped	40	16.9%
27	LTAP/OD at Signalized Junctions	6	2.5%
28	Vehicle Turning Right at Signalized Junctions	2	0.8%
29	LTAP/OD at Non-Signalized Junctions	3	1.3%
30	Straight Crossing Paths at Non-Signalized Junctions	7	3.0%
31	Vehicle(s) Turning at Non-Signalized Junctions	10	4.2%
32	Evasive Action With Prior Vehicle Maneuver	5	2.1%
33	Evasive Action Without Prior Vehicle Maneuver	3	1.3%
34	Non-Collision Incident	1	0.4%
35	Object Crash With Prior Vehicle Maneuver	6	2.5%
36	Object Crash Without Prior Vehicle Maneuver	5	2.1%
37	Other: Hit-and-Run (7 cases); On-Road Rollover (3 cases); No Driver Present (1 case); Other (1 case)	12	5.1%
	Total	236	100.0%

5.2. Mapping of 44 Crashes

Table 17 maps the 44 crashes to this new pre-crash scenario typology. Most of the 44 crashes are represented either directly or indirectly by the different variations of pre-crash scenarios in the new typology. For example, number 37 addresses emergency vehicles as they pass through signalized intersections on red. This crash is assigned to "running red light" scenario in the new typology even though the analysis of light-vehicle crashes in this report excludes emergency vehicles. However, the GES contains the needed variables to explicitly describe emergency-vehicle crashes that involve police cars, ambulances, or firefighting vehicles. Moreover, number 101 (new crash due to new safety technology) is assigned to "other" since it is not practical at this time to quantify this crash using existing national crash databases. Other crash numbers, such as 52 (tailgate), 61 (pedal miss), and 64 (stutter stop), are classified, respectively, under lead vehicle decelerating, stopped, and accelerating due to the lack of GES variables and codes that refer to these particular events. As seen in Table 17, there are 11 pre-crash scenarios in the new typology, accounting for about 10 percent of all light-vehicle crashes, which do not match any of the 44 crashes.

Table 17. Mapping of 44 Crashes to New Pre-Crash Scenario Typology

No.	New Crash Typology	44 Crashes
1	Vehicle Failure	68
2	Control Loss With Prior Vehicle Action	10
3	Control Loss Without Prior Vehicle Action	11, 12, 18, 91
4	Running Red Light	22, 37, 94
5	Running Stop Sign	28, 30
6	Road Edge Departure With Prior Vehicle Maneuver	10
7	Road Edge Departure Without Prior Vehicle Maneuver	9, 18
8	Road Edge Departure While Backing Up	19
9	Animal Crash With Prior Vehicle Maneuver	3
10	Animal Crash Without Prior Vehicle Maneuver	3
11	Pedestrian Crash With Prior Vehicle Maneuver	1
12	Pedestrian Crash Without Prior Vehicle Maneuver	1
13	Pedalcyclist Crash With Prior Vehicle Maneuver	
14	Pedalcyclist Crash Without Prior Vehicle Maneuver	
15	Backing Up Into Another Vehicle	48, 82
16	Vehicle(s) Turning – Same Direction	47, 83
17	Vehicle(s) Parking – Same Direction	
18	Vehicle(s) Changing Lanes – Same Direction	75, 76, 79, 80
19	Vehicle(s) Drifting – Same Direction	
20	Vehicle(s) Making a Maneuver – Opposite Direction	
21	Vehicle(s) Not Making a Maneuver – Opposite Direction	91, 92, 93
22	Following Vehicle Making a Maneuver	58
23	Lead Vehicle Accelerating	64
24	Lead Vehicle Moving at Lower Constant Speed	
25	Lead Vehicle Decelerating	52, 62, 74, 78
26	Lead Vehicle Stopped	56, 61, 62, 66
27	LTAP/OD at Signalized Junctions	96, 99
28	Vehicle Turning Right at Signalized Junctions	
29	LTAP/OD at Non-Signalized Junctions	96, 99
30	Straight Crossing Paths at Non-Signalized Junctions	33
31	Vehicle(s) Turning at Non-Signalized Junctions	35, 38, 40, 44
32	Evasive Action With Prior Vehicle Maneuver	
33	Evasive Action Without Prior Vehicle Maneuver	13
34	Non-Collision Incident	
35	Object Crash With Prior Vehicle Maneuver	
36	Object Crash Without Prior Vehicle Maneuver	
37	Other	100, 101

5.3. Mapping of Crash Types

Table shows an approximate mapping of pre-crash scenarios in the new typology to the eleven crash types identified in prior NHTSA studies. This is an approximation because some of these pre-crash scenarios can lead to different crash types. These eleven crash types are defined as follows:

- Rear-End: The front of a following vehicle strikes the rear of a lead vehicle, both traveling in the same direction.
- Crossing Paths: One moving vehicle cuts across the path of another, initially approaching from either lateral or opposite directions, in such a way that they collide at or near a junction.
- Run-Off-Road: The first harmful event occurs off the roadway after a vehicle in transport departs the travel portion of the roadway.
- Lane Change: A vehicle attempts to change lanes, merge, pass, leave/enter a parking position, or drift and strikes or is struck by another vehicle in the adjacent lane, both traveling in the same direction.
- Animal: A moving vehicle collides with an animal.
- Opposite Direction: A vehicle strikes another vehicle in the adjacent lane, traveling in the opposite direction, resulting in a frontal or sideswipe impact.
- Backing: A vehicle strikes or is struck by an obstacle or another vehicle while moving backwards.
- Pedestrian: A moving vehicle collides with a pedestrian.
- Pedalcyclist: A vehicle strikes or is struck by a pedalcyclist.
- Object: A vehicle strikes an object on the road.
- Other: This type encompasses the remaining crashes that are coded as "Other", "Unknown", or "No Impact" (e.g., fire or immersion) in the Accident Type variable.

Table 18. Mapping of Crash Types to New Pre-Crash Scenario Typology

No.	Pre-Crash Scenario	Crash Type
1	Vehicle Failure	Run-Off-Road
2	Control Loss With Prior Vehicle Action	
3	Control Loss Without Prior Vehicle Action	
4	Running Red Light	Crossing Paths
5	Running Stop Sign	
6	Road Edge Departure With Prior Vehicle Maneuver	Run-Off-Road
7	Road Edge Departure Without Prior Vehicle Maneuver	
8	Road Edge Departure While Backing Up	
9	Animal Crash With Prior Vehicle Maneuver	Animal
10	Animal Crash Without Prior Vehicle Maneuver	
11	Pedestrian Crash With Prior Vehicle Maneuver	Pedestrian
12	Pedestrian Crash Without Prior Vehicle Maneuver	
13	Pedalcyclist Crash With Prior Vehicle Maneuver	Pedalcyclist
14	Pedalcyclist Crash Without Prior Vehicle Maneuver	
15	Backing Up Into Another Vehicle	Backing
16	Vehicle(s) Turning – Same Direction	Lane Change
17	Vehicle(s) Parking – Same Direction	
18	Vehicle(s) Changing Lanes – Same Direction	
19	Vehicle(s) Drifting – Same Direction	
20	Vehicle(s) Making a Maneuver – Opposite Direction	Opposite Direction
21	Vehicle(s) Not Making a Maneuver – Opposite Direction	
22	Following Vehicle Making a Maneuver	Rear-End
23	Lead Vehicle Accelerating	
24	Lead Vehicle Moving at Lower Constant Speed	
25	Lead Vehicle Decelerating	
26	Lead Vehicle Stopped	
27	LTAP/OD at Signalized Junctions	Crossing Paths
28	Vehicle Turning Right at Signalized Junctions	
29	LTAP/OD at Non-Signalized Junctions	
30	Straight Crossing Paths at Non-Signalized Junctions	
31	Vehicle(s) Turning at Non-Signalized Junctions	
32	Evasive Action With Prior Vehicle Maneuver	Run-Off-Road
33	Evasive Action Without Prior Vehicle Maneuver	
34	Non-Collision Incident	Other
35	Object Crash With Prior Vehicle Maneuver	Object
36	Object Crash Without Prior Vehicle Maneuver	
37	Other	Other

6. CONCLUSIONS

This report defined and statistically described a novel typology of pre-crash scenarios representing all light-vehicle crashes based on 2004 GES statistics. These pre-crash scenarios depict vehicle movements and dynamics as well as the critical event that occur immediately before impact in a crash. This report quantified the severity of these scenarios and portrayed them by crash contributing factors and circumstances in terms of the driving environment, driver, and vehicle. This typology establishes a common vehicle safety research foundation for public and private organizations, which will serve as a tool to identify intervention opportunities, set research priorities and direction in technology development, and evaluate the effectiveness of selected crash countermeasure systems. It also provides a consistent crash problem definition for developers of crash avoidance technologies, simplifies crash characteristics for system designers, and prevents double counting of system safety benefits.

This new typology consists of 37 pre-crash scenarios (including "other") that accounted for approximately 5,942,000 police-reported crashes involving at least one light vehicle. These crashes resulted in an estimated economic cost of $120 billion and 2,767,000 functional years lost. These statistics do not incorporate data from non-police-reported crashes. Excluding "other" scenario, this new pre-crash scenario typology represents about 99.4 percent of all light-vehicle crashes. This typology is nationally representative and can be updated on an annual basis using GES and CDS crash databases.

Pre-crash scenarios of this new typology were ranked using three measures: crash frequency, economic cost, and functional years lost. The following dominant scenarios emerged using the top five scenarios in each of the three measures:

1. Control loss without prior vehicle action
2. Lead vehicle stopped
3. Road edge departure without prior vehicle maneuver
4. Vehicle(s) turning at non-signalized junctions
5. Straight crossing paths at non-signalized junctions
6. Lead vehicle decelerating
7. Vehicle(s) not making a maneuver – opposite direction

Despite its limitations, GES remains the best available source to identify nationally representative, dynamically distinct pre-crash scenarios. Moreover, GES contains a multitude of variables that allow the statistical description of driving circumstances at the time of the crash, driver contributing factors, and vehicle conditions. It is noteworthy that GES underestimates some crash scenarios or contributing factors due to the lack of information or non-specific information in police collision reports.

Crash statistics of this new typology should be updated on an annual basis using GES or CDS so as to ensure the consistency of its scenario ranking and national representativeness of all light-vehicle crashes over time. Such updates also serve to identify trends in crash statistics and assess effectiveness of new automotive safety technologies in the vehicle fleet such as electronic stability control systems.

Some hot-deck imputed GES variables were used to derive counts of crash frequency. It is recommended for further analysis that the percentage distribution between the original and the hot-deck variables be examined to assess any significant difference between the two sets of variables. If any significant difference existed, then further investigation might be necessary to determine which variables are more appropriate to be used for crash frequency counts.

7. REFERENCES

[1] NAO Engineering, Safety & Restraints Center, Crash Avoidance Department, "*44-Crashes*", General Motors Corporation, Version 3.0, January 1997.

[2] Crash Avoidance Metrics Partnership, "Enhanced Digital Mapping Project – Final Report". U.S. Department of Transportation, National Highway Traffic Safety Administration, November 2004.

[3] W.G. Najm, B. Sen, J.D. Smith, and B.N. Campbell, "*Analysis of Light Vehicle Crashes and Pre-Crash Scenarios Based on the 2000 General Estimates System*". DOT-VNTSC-NHTSA-02-04, DOT HS 809 573, February 2003.

[4] National Center for Statistics and Analysis, "*National Automotive Sampling System (NASS) General Estimates System (GES) Analytical User's Manual 1988-2004*". U.S. Department of Transportation, National Highway Traffic Safety Administration, 2005.

[5] D. Willke, S. Summers, J. Wang, J. Lee, S. Partyka, and S. Duffy, "*Ejection Mitigation Using Advanced Glazing: Status Report II*", Transportation Research Center, August 1999.

[6] L. Blincoe, A. Seay, E. Zaloshnja, T. Miller, E. Romano, S. Luchter, and R. Spicer, "*The Economic Impact of Motor Vehicle Crashes 2000*". U.S. Department of Transportation, National Highway Traffic Safety Administration, May 2002.

[7] T. Miller, J. Viner, S. Rossman, N. Pindus, W. Gellert, J. Douglass, A. Dillingham, and G. Blomquist, "*The Costs of Highway Crashes*". FHWA-RD-91-055, October 1991.

[8] W.G. Najm, M.D. Stearns, H. Howarth, J. Koopmann, and J. Hitz, "*Evaluation of an Automotive Rear-End Collision Avoidance System*". DOT-VNTSC-NHTSA-06-01, DOT HS 810 569, March 2006.

[9] Federal Highway Administration. 1995 Nationwide Personal Transportation Survey, www.bts.gov/ntda/npts/.

APPENDIX A. IDENTIFICATION CODES OF PRE-CRASH SCENARIOS USING THE GENERAL ESTIMATES SYSTEM

No.	Scenario	Single-Vehicle Crashes (VEH_INVL = 1)	Multi-Vehicle Crashes (VEH_INVL >= 2), First Event
1	No driver present	MANEUV_I = 0	
2	Vehicle failure	P_CRASH2 = 1 - 4	P_CRASH2 = 1 - 4 (at least one vehicle)
3	Control loss/vehicle action	P_CRASH2 = 5 - 9 AND MANEUV_I = 2 - 4, 6, 8 - 13, 15 - 97	Vx_P_CRASH2 = 5 - 9 AND Vx_MANEUV_I = 2 - 4, 6, 8 - 13, 15 - 97
		ACC_TYPE = 2, 7 AND MANEUV_I = 2 - 4, 6, 8 - 13, 15 - 97	Vx_ACC_TYPE = 34, 36, 54, 56 AND Vx_MANEUV_I = 2 - 4, 6, 8 - 13, 15 - 97
			Vx_ACC_TYPE = 2, 7 AND Vx_MANEUV_I = 2 - 4, 6, 8 - 13, 15 - 97
4	Control loss/no vehicle action	P_CRASH2 = 5 - 9 AND MANEUV_I = 1, 14	Vx_P_CRASH2 = 5 - 9 AND Vx_MANEUV_I = 1, 14
		ACC_TYPE = 2, 7 AND MANEUV_I = 1, 14	Vx_ACC_TYPE = 34, 36, 54, 56 AND Vx_MANEUV_I = 1, 14
			Vx_ACC_TYPE = 2, 7 AND Vx_MANEUV_I = 1, 14
5	Running red light	TRAF_CON = 1, 4 AND MVIOLATN = 7	TRAF_CON = 1 AND ACC_TYPE = 76, 77, 82, 83, 86 - 91
			TRAF_CON = 1, 4 AND MVIOLATN = 7
6	Running stop sign	TRAF_CON = 21 AND MVIOLATN = 7	TRAF_CON = 21 AND MVIOLATN = 7
7	Road edge departure/maneuver	P_CRASH2 = 10 - 14 AND MANEUV_I = 6, 8 - 12, 15 - 97	Vx_ACC_TYPE = 1, 6, 14 AND Vx_MANEUV_I = 6, 8 - 12, 15 - 97
		ACC_TYPE = 1, 6, 14 AND MANEUV_I = 6, 8 - 12, 15 - 97	
8	Road edge departure/no maneuver	P_CRASH2 = 10 - 14 AND MANEUV_I = 1 - 5, 7, 14	Vx_ACC_TYPE = 1, 6, 14 AND Vx_MANEUV_I = 1 - 5, 7, 14
		ACC_TYPE = 1, 6, 14 AND MANEUV_I = 1 - 5, 7, 14	
9	Road edge departure/backing	P_CRASH2 = 10 - 14 AND MANEUV_I = 13	Vx_ACC_TYPE = 1, 6, 14 AND Vx_MANEUV_I = 13
		ACC_TYPE = 1, 6, 14 AND MANEUV_I = 13	
		ACC_TYPE = 92	
10	Animal/maneuver	EVENT1_I = 24 AND MANEUV_I = 6, 8 - 13, 15 - 97	Vx_P_CRASH2 = 87 - 89 AND Vx_MANEUV_I = 6, 8 - 13, 15 - 97
		P_CRASH2 = 87 - 89 AND MANEUV_I = 6, 8 - 13, 15 - 97	EVENTNUM = 1 AND VEHNUM = x AND OBJCONT = 124 AND Vx_MANEUV_I = 6, 8 - 13, 15 - 97
11	Animal/no maneuver	EVENT1_I = 24 AND MANEUV_I = 1 - 5, 7, 14	Vx_P_CRASH2 = 87 - 89 AND Vx_MANEUV_I = 1 - 5, 7, 14
		P_CRASH2 = 87 - 89 AND MANEUV_I = 1 - 5, 7, 14	EVENTNUM = 1 AND VEHNUM = x AND OBJCONT = 124 AND Vx_MANEUV_I = 1 - 5, 7, 14
12	Pedestrian/maneuver	EVENT1_I = 21 AND MANEUV_I = 6, 8 - 13, 15 - 97	Vx_P_CRASH2 = 80 - 82 AND Vx_MANEUV_I = 6, 8 - 13, 15 - 97
		P_CRASH2 = 80 - 82 AND MANEUV_I = 6, 8 - 13, 15 - 97	EVENTNUM = 1 AND VEHNUM = x AND OBJCONT = 121 AND Vx_MANEUV_I = 6, 8 - 13, 15 - 97
13	Pedestrian/no maneuver	EVENT1_I = 21 AND MANEUV_I = 1 - 5, 7, 14	Vx_P_CRASH2 = 80 - 82 AND Vx_MANEUV_I = 1 - 5, 7, 14
		P_CRASH2 = 80 - 82 AND MANEUV_I = 1 - 5, 7, 14	EVENTNUM = 1 AND VEHNUM = x AND OBJCONT = 121 AND Vx_MANEUV_I = 1 - 5, 7, 14
14	Cyclist/maneuver	EVENT1_I = 22 AND MANEUV_I = 6, 8 - 13, 15 - 97	Vx_P_CRASH2 = 83 - 85 AND Vx_MANEUV_I = 6, 8 - 13, 15 - 97

No.	Scenario	Single-Vehicle Crashes (VEH_INVL = 1)	Multi-Vehicle Crashes (VEH_INVL >= 2), First Event
15	Cyclist/no maneuver	P_CRASH2 = 83 - 85 AND MANEUV_1 = 6, 8 - 13, 15 - 97	EVENTNUM = 1 AND VEHNUM = x AND OBJCONT = 122 AND Vx_MANEUV_1 = 6, 8 - 13, 15 - 97
		EVENT1_1 = 22 AND MANEUV_1 = 1 - 5, 7, 14	Vx_P_CRASH2 = 83 - 85 AND Vx_MANEUV_1 = 1 - 5, 7, 14
		P_CRASH2 = 83 - 85 AND MANEUV_1 = 1 - 5, 7, 14	EVENTNUM = 1 AND VEHNUM = x AND OBJCONT = 122 AND Vx_MANEUV_1 = 1 - 5, 7, 14
16	Backing into vehicle	P_CRASH2 = 56	ACC_TYPE = 92, 93 AND EVENT1_1 = 25
17	Turning/same direction		ACC_TYPE = 44 - 49, 70 - 73 AND MANEUV_1 = 10 - 12
			ACC_TYPE = 20 - 43 AND Vx_VROLE_1 = 2 AND Vx_MANEUV_1 = 10 - 12
			MANEUV_1 = 10 -12 AND P_CRASH2 = 60, 61
18	Parking/same direction	P_CRASH2 = 64	ACC_TYPE = 44 - 49, 70 - 73 AND MANEUV_1 = 8, 9
			ACC_TYPE = 20 - 43 AND Vx_VROLE_1 = 2 AND Vx_MANEUV_1 = 8, 9
			MANEUV_1 = 8, 9 AND P_CRASH2 = 60, 61
			P_CRASH2 = 64
19	Changing lanes/same direction	P_CRASH2 = 60, 61	ACC_TYPE = 44 - 49, 70 - 73 AND MANEUV_1 = 6, 15, 16
			ACC_TYPE = 20 - 43 AND Vx_VROLE_1 = 2 AND Vx_MANEUV_1 = 6, 15, 16
			MANEUV_1 = 6, 15, 16 AND P_CRASH2 = 60, 61
20	Drifting/same lane		ACC_TYPE = 44 - 49, 70 - 73 AND MANEUV_1 = 1 - 5, 7, 14
			ACC_TYPE = 20 - 43 AND Vx_VROLE_1 = 2 AND Vx_P_CRASH2 = 10, 11
21	Opposite direction/maneuver	P_CRASH2 = 54, 62, 63 AND MANEUV_1 = 6, 8 - 13, 15 - 97	ACC_TYPE = 50 - 67 AND MANEUV_1 = 6, 8 - 13, 15 - 97
22	Opposite direction/no maneuver	P_CRASH2 = 54, 62, 63 AND MANEUV_1 = 1 - 5, 7, 14	ACC_TYPE = 50 - 67 AND MANEUV_1 = 1 - 5, 7, 14
23	Rear-end/striking maneuver	P_CRASH2 = 50 - 52 AND MANEUV_1 = 6, 8 - 13, 15 - 97	ACC_TYPE = 20 - 43 AND Vx_VROLE_1 = 1 AND Vx_MANEUV_1 = 6, 8 - 13, 15 - 97
			Vx_VROLE_1 = 1 AND Vx_MANEUV_1 = 6, 8 - 13, 15 - 97 AND Vx_P_CRASH2 = 50, 51, 52
24	Rear-end/LVA		ACC_TYPE = 20 - 43 AND Vx_VROLE_1 = 2 AND Vx_MANEUV_1 = 3, 4
			Vx_MANEUV_1 = 3, 4 AND Vx_P_CRASH2 = 53
25	Rear-end/LVM	P_CRASH2 = 51	ACC_TYPE = 25 - 27
			ACC_TYPE = 20 - 43 AND Vx_VROLE_1 = 2 AND Vx_MANEUV_1 = 1, 14
			ACC_TYPE = 20 - 43 AND Vx_VROLE_1 = 1 AND Vx_P_CRASH2 = 51
			P_CRASH2 = 51
			Vx_MANEUV_1 = 1, 14 AND Vx_P_CRASH2 = 53
26	Rear-end/LVD	P_CRASH2 = 52	ACC_TYPE = 29 - 31
			ACC_TYPE = 20 - 43 AND Vx_VROLE_1 = 2 AND Vx_MANEUV_1 = 2
			ACC_TYPE = 20 - 43 AND Vx_VROLE_1 = 1 AND Vx_P_CRASH2 = 52

No.	Scenario	Single-Vehicle Crashes (VEH_INVL = 1)	Multi-Vehicle Crashes (VEH_INVL >= 2), First Event
			P_CRASH2 = 52
27	Rear-end/LVS	P_CRASH2 = 50	Vx_MANEUV_1 = 2 AND Vx_P_CRASH2 = 53
			ACC_TYPE = 21 - 23
			ACC_TYPE = 20 - 43 AND Vx_VROLE_1 = 2 AND Vx_MANEUV_1 = 5, 7
			ACC_TYPE = 20 - 43 AND Vx_VROLE_1 = 1 AND Vx_P_CRASH2 = 50
			P_CRASH2 = 50
			Vx_MANEUV_1 = 5, 7 AND Vx_P_CRASH2 = 53
			ACC_TYPE = 20 - 43 AND Vx_MANEUV_1 = 1 AND Vy_MANEUV_1 = 0
28	LTAP/OD @ signal		ACC_TYPE = 1 AND ACC_TYPE = 68, 69
			TRAF_CON = 1 AND MANEUV_1 = 11 AND P_CRASH2 = 54, 62, 63
			TRAF_CON = 1 AND Vx_P_CRASH2 = 15 AND Vy_P_CRASH2 = 54, 62, 63
			TRAF_CON = 1 AND Vx_MANEUV_1 = 11 AND Vy_MANEUV_1 not 10 AND ACC_TYPE = 74, 75
29	Turn right @ signal		TRAF_CON = 1 AND ACC_TYPE = 78 - 81
			TRAF_CON = 1 AND MANEUV_1 = 10 AND P_CRASH2 = 65 - 68
			TRAF_CON = 1 AND Vx_P_CRASH2 = 16 AND Vy_P_CRASH2 = 65 - 68
			TRAF_CON = 1 AND V_MANEUV_1 = 10 AND ACC_TYPE = 74, 75, 84, 85
30	LTAP/OD @ non signal		TRAF_CON not 1 AND ACC_TYPE = 68, 69
			TRAF_CON not 1 AND MANEUV_1 = 11 AND P_CRASH2 = 54, 62, 63
			TRAF_CON not 1 AND Vx_P_CRASH2 = 15 AND Vy_P_CRASH2 = 54, 62, 63
31	SCP @ non signal	TRAF_CON not 1 AND P_CRASH2 = 66, 71	TRAF_CON not 1 AND ACC_TYPE = 86 - 91
			TRAF_CON not 1 AND MANEUV_1 not 10-12 AND P_CRASH2 = 65 - 68, 70 - 78
			TRAF_CON not 1 AND Vx_P_CRASH2 not 15, 16 AND Vy_P_CRASH2 = 65 - 68, 70 - 78
32	Turn @ non signal	TRAF_CON not 1 AND P_CRASH2 = 65, 67, 68, 70, 72, 73	TRAF_CON not 1 AND ACC_TYPE = 74 - 85
			TRAF_CON not 1 AND MANEUV_1 = 10-12 AND P_CRASH2 = 65 - 68, 70 - 78
			TRAF_CON not 1 AND Vx_P_CRASH2 = 15, 16 AND Vy_P_CRASH2 = 65 - 68, 70 - 78
33	Avoidance/maneuver	ACC_TYPE = 3, 8 AND MANEUV_1 = 6, 8 - 13, 15 - 97	
		P_CRASH2 = 50 - 78 AND MANEUV_1 = 6, 8 - 13, 15 - 97	
34	Avoidance/no maneuver	ACC_TYPE = 3, 8 AND MANEUV_1 = 1 - 5, 7, 14	
		P_CRASH2 = 50 - 78 AND MANEUV_1 = 1 - 5, 7, 14	

No.	Scenario	Single-Vehicle Crashes (VEH_INVL = 1)	Multi-Vehicle Crashes (VEH_INVL >= 2), First Event
35	Rollover	ROLLOVER=10 OR EVENT1_1=1	
36	Noncollision - No Impact	EVENT1_1=2 - 10	
		ACC_TYPE = 00	
37	Object/maneuver	P_CRASH2 = 90, 91, 92 AND MANEUV_1 = 6, 8 - 13, 15 - 97	
		ACC_TYPE = 12 AND MANEUV_1 = 6, 8 - 13, 15 - 97	
		ACC_TYPE = 11 AND MANEUV_1 = 6, 8 - 13, 15 - 97	
		EVENT1_1 = 21 - 29, 31 - 59 AND MANEUV_1 = 6, 8 - 13, 15 - 97	
38	Object/no maneuver	P_CRASH2 = 90, 91, 92 AND MANEUV_1 = 1 - 5, 7, 14	
		ACC_TYPE = 12 AND MANEUV_1 = 1 - 5, 7, 14	
		ACC_TYPE = 11 AND MANEUV_1 = 1 - 5, 7, 14	
		EVENT1_1 = 21 - 29, 31 - 59 AND MANEUV_1 = 1 - 5, 7, 14	
39	Hit-and-run	HITRUN_1 = 1	
40	Other - Rear-End		ACC_TYPE = 20 - 43
41	Other - Sideswipe		ACC_TYPE = 44 - 49
42	Other - Opposite Direction		ACC_TYPE = 50 - 67
43	Other - Turn Across Path		ACC_TYPE = 68 - 75
44	Other - Turn Into Path		ACC_TYPE = 76 - 85
45	Other - Straight Paths		ACC_TYPE = 86 - 91
46	Other		

APPENDIX B. CRASH CHARACTERISTICS OF PRE-CRASH SCENARIOS

Vehicle Failure

Driving Environment

Category	Value	%
Lighting	Daylight	67%
	Dark Lighted	12%
	Dark	15%
	Dawn/Dusk	6%
Weather	Clear	87%
	Adverse	13%
Road Surface	Dry	83%
	Wet/Slippery	17%
Road Alignment	Straight	76%
	Curve	24%
Road Profile	Level	71%
	Other	29%
Land Use	Rural	64%
	Urban	36%
Day	Weekday	73%
	Weekend	27%
Relation to Roadway	On Roadway	27%
	Shoulder/Parking Lane	5%
	Off Roadway	67%
	Left Turn Lane	-
	Unknown	1%
Relation to Junction	Non-Junction	81%
	Intersection	4%
	Intersection-Related	9%
	Driveway/Alley	2%
	Entrance/Exit Ramp	3%
	Rail Grade Crossing	1%
	Other/Unknown	1%
Posted Speed Limit (mph)	<= 20	1%
	25	8%
	30	5%
	35	10%
	40	5%
	45	13%
	50	3%
	>= 55	55%
Traffic Control Device	No Traffic Controls	81%
	Traffic Signal	9%
	Stop/Yield Sign	3%
	Other	8%

Driver

Category	Value	%
Alcohol	Yes	2%
	No	98%
Vision Obscured	No Obstruction	83%
	Vision Obscured	1%
	Unknown	16%
Driver Distracted	Inattention	2%
	Sleepy	0.01%
	Not Distracted	54%
	Unknown	43%
Speeding	Yes	8%
	No	91%
	Unknown	1%
Violation	Speeding	-
	Reckless	1%
	None	77%
	Other	22%
	Unknown	1%
Impairment	Ill/Blackout	-
	Drowsy	0.01%
	None	97%
	Other	1%
	Unknown	2%
Gender	Male	64%
	Female	36%
Age	Younger <= 24	39%
	Middle = 25 to 64	57%
	Older >= 65	4%

Vehicle

Category	Value	%
Contributing Factors	Yes	99%
	No	-
	Unknown	1%
Rollover	Yes	22%
	No	78%
Pre-Event Movement	No Driver Present	-
	Going Straight	75%
	Decelerating in Traffic Lane	2%
	Accelerating in Traffic Lane	0.2%
	Starting in Traffic Lane	0.2%
	Stopped in Traffic Lane	1%
	Passing Another Vehicle	1%
	Parked in Travel Lane	1%
	Leaving a Parked Position	0.02%
	Entering a Parked Position	-
	Turning Right	2%
	Turning Left	4%
	Making U-turn	-
	Backing Up	1%
	Negotiating a Curve	10%
	Changing Lanes	1%
	Merging	0.2%
	Prior Corrective Action	0.1%
	Other	2%
Driver Avoidance Maneuver	Object in Road	-
	Poor Road Conditions	0.1%
	Animal in Road	-
	Vehicle in Road	1%
	Non-Motorist in Road	-
	Hit and Run	2%
	No Driver Present	-
	Other Avoidance Maneuver	-
	Unknown	55%
	None	41%
	Phantom Vehicle	-
Corrective Action Attempted	No Driver Present	-
	No Avoidance Maneuver	18%
	Braking	6%
	Releasing Brakes	-
	Steering	7%
	Braked and Steered	2%
	Accelerated	0.02%
	Accelerated and Steered	-
	Other	2%
	Unknown	67%

Driver and vehicle statistics represent the light vehicle with a component failure.

Control Loss With Prior Vehicle Action

Driving Environment

Lighting	Daylight	54%
	Dark Lighted	24%
	Dark	16%
	Dawn/Dusk	6%
Weather	Clear	60%
	Adverse	40%
Road Surface	Dry	41%
	Wet/Slippery	59%
Road Alignment	Straight	87%
	Curve	13%
Road Profile	Level	73%
	Other	27%
Land Use	Rural	53%
	Urban	47%
Day	Weekday	71%
	Weekend	29%
Relation to Roadway	On Roadway	28%
	Shoulder/Parking Lane	4%
	Off Roadway	68%
	Left Turn Lane	-
	Unknown	0.2%
Relation to Junction	Non-Junction	34%
	Intersection	5%
	Intersection-Related	45%
	Driveway/Alley	7%
	Entrance/Exit Ramp	7%
	Rail Grade Crossing	0.4%
	Other/Unknown	2%
Posted Speed Limit (mph)	<= 20	2%
	25	16%
	30	10%
	35	15%
	40	6%
	45	13%
	50	4%
	>= 55	34%
Traffic Control Device	No Traffic Controls	67%
	Traffic Signal	14%
	Stop/Yield Sign	11%
	Other	7%

Driver

Alcohol	Yes	13%
	No	87%
Vision Obscured	No Obstruction	66%
	Vision Obscured	1%
	Unknown	33%
Driver Distracted	Inattention	11%
	Sleepy	0.1%
	Not Distracted	45%
	Unknown	44%
Speeding	Yes	56%
	No	41%
	Unknown	3%
Violation	Speeding	0.2%
	Reckless	2%
	None	49%
	Other	43%
	Unknown	6%
Impairment	Ill/Blackout	1%
	Drowsy	0.4%
	None	86%
	Other	8%
	Unknown	5%
Gender	Male	65%
	Female	35%
Age	Younger <= 24	53%
	Middle = 25 to 64	43%
	Older >= 65	3%

Vehicle

Contributing Factors	Yes	2%
	No	85%
	Unknown	13%
Rollover	Yes	12%
	No	88%
Pre-Event Movement	No Driver Present	-
	Going Straight	-
	Decelerating in Traffic Lane	14%
	Accelerating in Traffic Lane	1%
	Starting in Traffic Lane	1%
	Stopped in Traffic Lane	-
	Passing Another Vehicle	8%
	Parked in Travel Lane	-
	Leaving a Parked Position	3%
	Entering a Parked Position	0.1%
	Turning Right	21%
	Turning Left	26%
	Making U-turn	1%
	Backing Up	1%
	Negotiating a Curve	-
	Changing Lanes	14%
	Merging	4%
	Prior Corrective Action	1%
	Other	5%
Driver Avoidance Maneuver	Object in Road	-
	Poor Road Conditions	0.4%
	Animal in Road	1%
	Vehicle in Road	6%
	Non-Motorist in Road	0.004%
	Hit and Run	12%
	No Driver Present	-
	Other Avoidance Maneuver	0.1%
	Unknown	52%
	None	29%
	Phantom Vehicle	1%
Corrective Action Attempted	No Driver Present	-
	No Avoidance Maneuver	12%
	Braking	8%
	Releasing Brakes	-
	Steering	7%
	Braked and Steered	1%
	Accelerated	1%
	Accelerated and Steered	0.2%
	Other	0.2%
	Unknown	72%

Driver and vehicle statistics represent the light vehicle that lost control.

Control Loss Without Prior Vehicle Action

Driving Environment

Category	Value	%
Lighting	Daylight	53%
	Dark Lighted	14%
	Dark	27%
	Dawn/Dusk	5%
Weather	Clear	56%
	Adverse	44%
Road Surface	Dry	38%
	Wet/Slippery	62%
Road Alignment	Straight	58%
	Curve	42%
Road Profile	Level	65%
	Other	35%
Land Use	Rural	66%
	Urban	34%
Day	Weekday	69%
	Weekend	31%
Relation to Roadway	On Roadway	11%
	Shoulder/Parking Lane	4%
	Off Roadway	85%
	Left Turn Lane	-
	Unknown	0.3%
Relation to Junction	Non-Junction	88%
	Intersection	0.5%
	Intersection-Related	4%
	Driveway/Alley	0.3%
	Entrance/Exit Ramp	4%
	Rail Grade Crossing	0.2%
	Other/Unknown	2%
Posted Speed Limit (mph)	<= 20	2%
	25	8%
	30	7%
	35	11%
	40	5%
	45	14%
	50	3%
	>= 55	50%
Traffic Control Device	No Traffic Controls	89%
	Traffic Signal	1%
	Stop/Yield Sign	1%
	Other	8%

Driver

Category	Value	%
Alcohol	Yes	12%
	No	88%
Vision Obscured	No Obstruction	70%
	Vision Obscured	2%
	Unknown	29%
Driver Distracted	Inattention	11%
	Sleepy	2%
	Not Distracted	44%
	Unknown	43%
Speeding	Yes	58%
	No	39%
	Unknown	2%
Violation	Speeding	0.2%
	Reckless	2%
	None	59%
	Other	35%
	Unknown	3%
Impairment	Ill/Blackout	2%
	Drowsy	2%
	None	83%
	Other	7%
	Unknown	6%
Gender	Male	61%
	Female	39%
Age	Younger <= 24	45%
	Middle = 25 to 64	52%
	Older >= 65	3%

Vehicle

Category	Value	%
Contributing Factors	Yes	2%
	No	90%
	Unknown	8%
Rollover	Yes	23%
	No	77%
Pre-Event Movement	No Driver Present	-
	Going Straight	65%
	Decelerating in Traffic Lane	-
	Accelerating in Traffic Lane	-
	Starting in Traffic Lane	-
	Stopped in Traffic Lane	-
	Passing Another Vehicle	-
	Parked in Travel Lane	-
	Leaving a Parked Position	-
	Entering a Parked Position	-
	Turning Right	-
	Turning Left	-
	Making U-turn	-
	Backing Up	-
	Negotiating a Curve	35%
	Changing Lanes	-
	Merging	-
	Prior Corrective Action	-
	Other	-
Driver Avoidance Maneuver	Object in Road	0.4%
	Poor Road Conditions	1%
	Animal in Road	1%
	Vehicle in Road	3%
	Non-Motorist in Road	0.03%
	Hit and Run	6%
	No Driver Present	-
	Other Avoidance Maneuver	0.1%
	Unknown	46%
	None	43%
	Phantom Vehicle	1%
Corrective Action Attempted	No Driver Present	-
	No Avoidance Maneuver	14%
	Braking	6%
	Releasing Brakes	0.03%
	Steering	11%
	Braked and Steered	1%
	Accelerated	0.1%
	Accelerated and Steered	0.02%
	Other	1%
	Unknown	67%

Driver and vehicle statistics represent the light vehicle that lost control.

Running Red Light

Driving Environment

Lighting	Daylight	75%
	Dark Lighted	19%
	Dark	3%
	Dawn/Dusk	3%
Weather	Clear	88%
	Adverse	12%
Road Surface	Dry	81%
	Wet/Slippery	19%
Road Alignment	Straight	94%
	Curve	6%
Road Profile	Level	82%
	Other	18%
Land Use	Rural	40%
	Urban	60%
Day	Weekday	75%
	Weekend	25%
Relation to Roadway	On Roadway	100%
	Shoulder/Parking Lane	-
	Off Roadway	0.2%
	Left Turn Lane	0.2%
	Unknown	-
Relation to Junction	Non-Junction	-
	Intersection	93%
	Intersection-Related	3%
	Driveway/Alley	2%
	Entrance/Exit Ramp	2%
	Rail Grade Crossing	0.05%
	Other/Unknown	1%
Posted Speed Limit (mph)	<= 20	1%
	25	6%
	30	13%
	35	33%
	40	15%
	45	22%
	50	4%
	>= 55	5%
Traffic Control Device	No Traffic Controls	-
	Traffic Signal	100%
	Stop/Yield Sign	-
	Other	-

Driver

Alcohol	Yes	4%
	No	96%
Vision Obscured	No Obstruction	71%
	Vision Obscured	3%
	Unknown	26%
Driver Distracted	Inattention	32%
	Sleepy	0.2%
	Not Distracted	37%
	Unknown	31%
Speeding	Yes	3%
	No	96%
	Unknown	1%
Violation	Speeding	0.1%
	Reckless	0.2%
	None	-
	Other	100%
	Unknown	-
Impairment	Ill/Blackout	0.2%
	Drowsy	0.2%
	None	96%
	Other	2%
	Unknown	1%
Gender	Male	53%
	Female	47%
Age	Younger <= 24	32%
	Middle = 25 to 64	55%
	Older >= 65	13%

Vehicle

Contributing Factors	Yes	1%
	No	95%
	Unknown	5%
Rollover	Yes	2%
	No	98%
Pre-Event Movement	No Driver Present	-
	Going Straight	85%
	Decelerating in Traffic Lane	2%
	Accelerating in Traffic Lane	0.05%
	Starting in Traffic Lane	1.8%
	Stopped in Traffic Lane	0.1%
	Passing Another Vehicle	1%
	Parked in Travel Lane	-
	Leaving a Parked Position	-
	Entering a Parked Position	-
	Turning Right	2%
	Turning Left	7%
	Making U-turn	0.03%
	Backing Up	-
	Negotiating a Curve	1%
	Changing Lanes	0.4%
	Merging	-
	Prior Corrective Action	-
	Other	0.5%
Driver Avoidance Maneuver	Object In Road	-
	Poor Road Conditions	-
	Animal In Road	-
	Vehicle In Road	7%
	Non-Motorist In Road	0.004%
	Hit and Run	2%
	No Driver Present	-
	Other Avoidance Maneuver	-
	Unknown	68%
	None	23%
	Phantom Vehicle	-
Corrective Action Attempted	No Driver Present	-
	No Avoidance Maneuver	21%
	Braking	6%
	Releasing Brakes	-
	Steering	3%
	Braked and Steered	-
	Accelerated	0.1%
	Accelerated and Steered	-
	Other	-
	Unknown	71%

Driver and vehicle statistics represent the violating light vehicle.

Running Stop Sign

Driving Environment

Lighting	Daylight	73%
	Dark Lighted	15%
	Dark	9%
	Dawn/Dusk	3%
Weather	Clear	88%
	Adverse	12%
Road Surface	Dry	83%
	Wet/Slippery	17%
Road Alignment	Straight	93%
	Curve	7%
Road Profile	Level	83%
	Other	17%
Land Use	Rural	60%
	Urban	40%
Day	Weekday	74%
	Weekend	26%
Relation to Roadway	On Roadway	92%
	Shoulder/Parking Lane	0.3%
	Off Roadway	8%
	Left Turn Lane	-
	Unknown	0.01%
Relation to Junction	Non-Junction	0.1%
	Intersection	91%
	Intersection-Related	8%
	Driveway/Alley	0.3%
	Entrance/Exit Ramp	0.4%
	Rail Grade Crossing	-
	Other/Unknown	1%
Posted Speed Limit (mph)	<= 20	2%
	25	29%
	30	18%
	35	20%
	40	7%
	45	9%
	50	1%
	>= 55	13%
Traffic Control Device	No Traffic Controls	-
	Traffic Signal	-
	Stop/Yield Sign	100%
	Other	-

Driver

Alcohol	Yes	7%
	No	93%
Vision Obscured	No Obstruction	72%
	Vision Obscured	4%
	Unknown	24%
Driver Distracted	Inattention	25%
	Sleepy	0.1%
	Not Distracted	33%
	Unknown	42%
Speeding	Yes	5%
	No	93%
	Unknown	2%
Violation	Speeding	-
	Reckless	0.4%
	None	-
	Other	100%
	Unknown	-
Impairment	Ill/Blackout	-
	Drowsy	0.1%
	None	92%
	Other	6%
	Unknown	2%
Gender	Male	61%
	Female	39%
Age	Younger <= 24	36%
	Middle = 25 to 64	50%
	Older >= 65	13%

Vehicle

Contributing Factors	Yes	1%
	No	93%
	Unknown	6%
Rollover	Yes	1%
	No	99%
Pre-Event Movement	No Driver Present	-
	Going Straight	76%
	Decelerating in Traffic Lane	1%
	Accelerating in Traffic Lane	-
	Starting in Traffic Lane	5%
	Stopped in Traffic Lane	0.1%
	Passing Another Vehicle	0.4%
	Parked in Travel Lane	-
	Leaving a Parked Position	0.1%
	Entering a Parked Position	-
	Turning Right	5%
	Turning Left	11%
	Making U-turn	-
	Backing Up	-
	Negotiating a Curve	0.1%
	Changing Lanes	-
	Merging	0.02%
	Prior Corrective Action	-
	Other	1%
Driver Avoidance Maneuver	Object in Road	-
	Poor Road Conditions	0.2%
	Animal in Road	0.2%
	Vehicle in Road	5%
	Non-Motorist in Road	-
	Hit and Run	3%
	No Driver Present	-
	Other Avoidance Maneuver	-
	Unknown	68%
	None	24%
	Phantom Vehicle	-
Corrective Action Attempted	No Driver Present	-
	No Avoidance Maneuver	18%
	Braking	6%
	Releasing Brakes	-
	Steering	1%
	Braked and Steered	0.5%
	Accelerated	0.004%
	Accelerated and Steered	0.02%
	Other	-
	Unknown	75%

Driver and vehicle statistics represent the violating light vehicle.

Road Edge Departure With Prior Vehicle Maneuver

Driving Environment *Driver* *Vehicle*

	Driving Environment			Driver			Vehicle	
Lighting	Daylight	45%	**Alcohol**	Yes	21%	**Contributing Factors**	Yes	1%
	Dark Lighted	32%		No	79%		No	69%
	Dark	17%	**Vision Obscured**	No Obstruction	47%		Unknown	30%
	Dawn/Dusk	6%		Vision Obscured	5%	**Rollover**	Yes	6%
Weather	Clear	89%		Unknown	48%		No	94%
	Adverse	11%	**Driver Distracted**	Inattention	23%	**Pre-Event Movement**	No Driver Present	-
Road Surface	Dry	84%		Sleepy	0.4%		Going Straight	-
	Wet/Slippery	16%		Not Distracted	21%		Decelerating in Traffic Lane	-
Road Alignment	Straight	86%		Unknown	55%		Accelerating in Traffic Lane	-
	Curve	14%	**Speeding**	Yes	14%		Starting in Traffic Lane	-
Road Profile	Level	81%		No	73%		Stopped in Traffic Lane	-
	Other	19%		Unknown	13%		Passing Another Vehicle	6%
Land Use	Rural	55%	**Violation**	Speeding	1%		Parked in Travel Lane	-
	Urban	45%		Reckless	2%		Leaving a Parked Position	5%
Day	Weekday	66%		None	41%		Entering a Parked Position	9%
	Weekend	34%		Other	38%		Turning Right	25%
Relation to Roadway	On Roadway	1%		Unknown	18%		Turning Left	28%
	Shoulder/Parking Lane	33%	**Impairment**	Ill/Blackout	1%		Making U-turn	1%
	Off Roadway	66%		Drowsy	1%		Backing Up	-
	Left Turn Lane	-		None	75%		Negotiating a Curve	-
	Unknown	-		Other	12%		Changing Lanes	9%
Relation to Junction	Non-Junction	35%		Unknown	11%		Merging	3%
	Intersection	-	**Gender**	Male	65%		Prior Corrective Action	3%
	Intersection-Related	50%		Female	35%		Other	11%
	Driveway/Alley	9%	**Age**	Younger <= 24	42%	**Driver Avoidance Maneuver**	Object in Road	0.01%
	Entrance/Exit Ramp	3%		Middle = 25 to 64	51%		Poor Road Conditions	0.02%
	Rail Grade Crossing	0.2%		Older >= 65	7%		Animal in Road	1%
	Other/Unknown	2%					Vehicle in Road	5%
Posted Speed Limit (mph)	<= 20	4%					Non-Motorist in Road	-
	25	35%					Hit and Run	28%
	30	10%					No Driver Present	-
	35	16%					Other Avoidance Maneuver	0.1%
	40	7%					Unknown	36%
	45	9%					None	29%
	50	3%					Phantom Vehicle	1%
	>= 55	16%				**Corrective Action Attempted**	No Driver Present	-
Traffic Control Device	No Traffic Controls	71%					No Avoidance Maneuver	19%
	Traffic Signal	9%					Braking	1%
	Stop/Yield Sign	10%					Releasing Brakes	-
	Other	9%					Steering	8%
							Braked and Steered	1%
							Accelerated	1%
							Accelerated and Steered	0.1%
							Other	1%
							Unknown	70%

Driver and vehicle statistics represent the light vehicle departing the road edge.

Road Edge Departure Without Prior Vehicle Maneuver

Driving Environment

Lighting	Daylight	44%
	Dark Lighted	25%
	Dark	28%
	Dawn/Dusk	3%
Weather	Clear	88%
	Adverse	12%
Road Surface	Dry	82%
	Wet/Slippery	18%
Road Alignment	Straight	74%
	Curve	26%
Road Profile	Level	76%
	Other	24%
Land Use	Rural	61%
	Urban	39%
Day	Weekday	63%
	Weekend	37%
Relation to Roadway	On Roadway	1%
	Shoulder/Parking Lane	27%
	Off Roadway	72%
	Left Turn Lane	-
	Unknown	0.1%
Relation to Junction	Non-Junction	89%
	Intersection	0.04%
	Intersection-Related	8%
	Driveway/Alley	0.2%
	Entrance/Exit Ramp	2%
	Rail Grade Crossing	0.1%
	Other/Unknown	1%
Posted Speed Limit (mph)	<= 20	3%
	25	22%
	30	10%
	35	16%
	40	6%
	45	12%
	50	3%
	>= 55	28%
Traffic Control Device	No Traffic Controls	86%
	Traffic Signal	2%
	Stop/Yield Sign	4%
	Other	8%

Driver

Alcohol	Yes	28%
	No	72%
Vision Obscured	No Obstruction	60%
	Vision Obscured	3%
	Unknown	37%
Driver Distracted	Inattention	27%
	Sleepy	12%
	Not Distracted	15%
	Unknown	46%
Speeding	Yes	16%
	No	74%
	Unknown	10%
Violation	Speeding	1%
	Reckless	2%
	None	43%
	Other	44%
	Unknown	10%
Impairment	Ill/Blackout	3%
	Drowsy	12%
	None	56%
	Other	19%
	Unknown	11%
Gender	Male	68%
	Female	32%
Age	Younger <= 24	41%
	Middle = 25 to 64	53%
	Older >= 65	6%

Vehicle

Contributing Factors	Yes	2%
	No	80%
	Unknown	18%
Rollover	Yes	12%
	No	88%
Pre-Event Movement	No Driver Present	-
	Going Straight	83%
	Decelerating in Traffic Lane	1%
	Accelerating in Traffic Lane	0.2%
	Starting in Traffic Lane	0.1%
	Stopped in Traffic Lane	-
	Passing Another Vehicle	-
	Parked in Travel Lane	-
	Leaving a Parked Position	-
	Entering a Parked Position	-
	Turning Right	-
	Turning Left	-
	Making U-turn	-
	Backing Up	-
	Negotiating a Curve	16%
	Changing Lanes	-
	Merging	-
	Prior Corrective Action	-
	Other	-
Driver Avoidance Maneuver	Object in Road	0.1%
	Poor Road Conditions	0.1%
	Animal in Road	1%
	Vehicle in Road	1%
	Non-Motorist in Road	0.01%
	Hit and Run	18%
	No Driver Present	-
	Other Avoidance Maneuver	0.1%
	Unknown	45%
	None	34%
	Phantom Vehicle	1%
Corrective Action Attempted	No Driver Present	-
	No Avoidance Maneuver	18%
	Braking	3%
	Releasing Brakes	-
	Steering	5%
	Braked and Steered	0.3%
	Accelerated	0.04%
	Accelerated and Steered	0.02%
	Other	0.2%
	Unknown	73%

Driver and vehicle statistics represent the light vehicle departing the road edge.

Road Edge Departure While Backing Up

Driving Environment

Lighting	Daylight	69%
	Dark Lighted	18%
	Dark	10%
	Dawn/Dusk	4%
Weather	Clear	93%
	Adverse	7%
Road Surface	Dry	85%
	Wet/Slippery	15%
Road Alignment	Straight	94%
	Curve	6%
Road Profile	Level	83%
	Other	17%
Land Use	Rural	49%
	Urban	51%
Day	Weekday	70%
	Weekend	30%
Relation to Roadway	On Roadway	5%
	Shoulder/Parking Lane	85%
	Off Roadway	10%
	Left Turn Lane	-
	Unknown	0.4%
Relation to Junction	Non-Junction	35%
	Intersection	0.5%
	Intersection-Related	3%
	Driveway/Alley	59%
	Entrance/Exit Ramp	-
	Rail Grade Crossing	0.1%
	Other/Unknown	3%
Posted Speed Limit (mph)	<= 20	31%
	25	46%
	30	9%
	35	8%
	40	1%
	45	1%
	50	1%
	>= 55	4%
Traffic Control Device	No Traffic Controls	92%
	Traffic Signal	1%
	Stop/Yield Sign	1%
	Other	6%

Driver

Alcohol	Yes	8%
	No	92%
Vision Obscured	No Obstruction	58%
	Vision Obscured	3%
	Unknown	39%
Driver Distracted	Inattention	32%
	Sleepy/Fell Asleep	0.02%
	Not Distracted	14%
	Unknown	53%
Speeding	Yes	1%
	No	87%
	Unknown	12%
Violation	Speeding	-
	Reckless	1%
	None	49%
	Other	32%
	Unknown	19%
Impairment	Ill/Blackout	-
	Sleepy/Drowsy	1%
	None	85%
	Other Impairment	3%
	Unknown	11%
Gender	Male	56%
	Female	44%
Age	Younger <= 24	34%
	Middle = 25 to 64	57%
	Older >= 65	9%

Vehicle

Contributing Factors	Yes	1%
	No	73%
	Unknown	26%
Rollover	Yes	1%
	No	99%
Pre-Event Movement	No Driver Present	-
	Going Straight	-
	Decelerating in Traffic Lane	-
	Accelerating in Traffic Lane	-
	Starting in Traffic Lane	-
	Stopped in Traffic Lane	-
	Passing Another Vehicle	-
	Parked in Travel Lane	-
	Leaving a Parked Position	9%
	Entering a Parked Position	3%
	Turning Right	-
	Turning Left	-
	Making U-turn	-
	Backing Up	87%
	Negotiating a Curve	-
	Changing Lanes	-
	Merging	-
	Prior Corrective Action	-
	Other	1%
Driver Avoidance Maneuver	Object in Road	-
	Poor Road Conditions	-
	Animal in Road	-
	Vehicle in Road	1%
	Non-Motorist in Road	-
	Hit and Run	24%
	No Driver Present	-
	Other Avoidance Maneuver	-
	Unknown	42%
	None	33%
	Phantom Vehicle	0.03%
Corrective Action Attempted	No Driver Present	-
	No Avoidance Maneuver	29%
	Braking	0.4%
	Releasing Brakes	-
	Steering	0.2%
	Braked and Steered	-
	Accelerated	0.4%
	Accelerated and Steered	-
	Other	0.2%
	Unknown	70%

Driver and vehicle statistics represent the backing light vehicle.

Animal Crash With Prior Vehicle Maneuver

Driving Environment

Lighting	Daylight	50%
	Dark Lighted	10%
	Dark	35%
	Dawn/Dusk	4%
Weather	Clear	87%
	Adverse	13%
Road Surface	Dry	41%
	Wet/Slippery	59%
Road Alignment	Straight	89%
	Curve	11%
Road Profile	Level	80%
	Other	20%
Land Use	Rural	79%
	Urban	21%
Day	Weekday	68%
	Weekend	32%
Relation to Roadway	On Roadway	83%
	Shoulder/Parking Lane	2%
	Off Roadway	14%
	Left Turn Lane	-
	Unknown	1%
Relation to Junction	Non-Junction	90%
	Intersection	-
	Intersection-Related	3%
	Driveway/Alley	1%
	Entrance/Exit Ramp	4%
	Rail Grade Crossing	-
	Other/Unknown	2%
Posted Speed Limit (mph)	<= 20	2%
	25	9%
	30	4%
	35	5%
	40	2%
	45	9%
	50	23%
	>= 55	46%
Traffic Control Device	No Traffic Controls	30%
	Traffic Signal	5%
	Stop/Yield Sign	-
	Other	65%

Driver

Alcohol	Yes	2%
	No	98%
Vision Obscured	No Obstruction	32%
	Vision Obscured	-
	Unknown	68%
Driver	Inattention	3%
	Sleepy	-
	Not Distracted	19%
	Unknown	77%
Speeding	Yes	1%
	No	87%
	Unknown	12%
Violation	Speeding	-
	Reckless	-
	None	95%
	Other	5%
	Unknown	-
Impairment	Ill/Blackout	-
	Sleepy/Drowsy	-
	None	94%
	Other	2%
	Unknown	4%
Gender	Male	50%
	Female	50%
Age	Younger <= 24	24%
	Middle = 25 to 64	70%
	Older >= 65	5%

Vehicle

Contributing Factors	Yes	-
	No	69%
	Unknown	31%
Rollover	Yes	5%
	No	95%
Pre-Event Movement	No Driver Present	-
	Going Straight	-
	Decelerating in Traffic Lane	-
	Accelerating in Traffic Lane	-
	Starting in Traffic Lane	-
	Stopped in Traffic Lane	-
	Passing Another Vehicle	6%
	Parked in Travel Lane	-
	Leaving a Parked Position	21%
	Entering a Parked Position	-
	Turning Right	1%
	Turning Left	1%
	Making U-turn	-
	Backing Up	-
	Negotiating a Curve	-
	Changing Lanes	3%
	Merging	-
	Prior Corrective Action	14%
	Other	53%
Driver Avoidance Maneuver	Object in Road	-
	Poor Road Conditions	-
	Animal in Road	19%
	Vehicle in Road	-
	Non-Motorist in Road	-
	Hit and Run	-
	No Driver Present	-
	Other Avoidance Maneuver	-
	Unknown	78%
	None	2%
	Phantom Vehicle	1%
Corrective Action Attempted	No Driver Present	-
	No Avoidance Maneuver	1%
	Braking	0.1%
	Releasing Brakes	-
	Steering	18%
	Braked and Steered	-
	Accelerated	-
	Accelerated and Steered	-
	Other	0.02%
	Unknown	81%

Animal Crash Without Prior Vehicle Maneuver

Driving Environment

Lighting	Daylight	24%
	Dark Lighted	8%
	Dark	58%
	Dawn/Dusk	9%
Weather	Clear	91%
	Adverse	9%
Road Surface	Dry	82%
	Wet/Slippery	18%
Road Alignment	Straight	89%
	Curve	11%
Road Profile	Level	74%
	Other	26%
Land Use	Rural	79%
	Urban	21%
Day	Weekday	70%
	Weekend	30%
Relation to Roadway	On Roadway	90%
	Shoulder/Parking Lane	0.4%
	Off Roadway	9%
	Left Turn Lane	-
	Unknown	0.1%
Relation to Junction	Non-Junction	97%
	Intersection	1%
	Intersection-Related	1%
	Driveway/Alley	-
	Entrance/Exit Ramp	1%
	Rail Grade Crossing	-
	Other/Unknown	1%
Posted Speed Limit (mph)	<= 20	1%
	25	5%
	30	2%
	35	8%
	40	4%
	45	12%
	50	5%
	>= 55	62%
Traffic Control Device	No Traffic Controls	91%
	Traffic Signal	1%
	Stop/Yield Sign	0.02%
	Other	8%

Driver

Alcohol	Yes	1%
	No	99%
Vision Obscured	No Obstruction	87%
	Vision Obscured	1%
	Unknown	13%
Driver Distracted	Inattention	1%
	Sleepy	-
	Not Distracted	74%
	Unknown	25%
Speeding	Yes	2%
	No	97%
	Unknown	1%
Violation	Speeding	-
	Reckless	0.1%
	None	97%
	Other	3%
	Unknown	0.1%
Impairment	Ill/Blackout	-
	Drowsy	-
	None	98%
	Other	0.3%
	Unknown	2%
Gender	Male	61%
	Female	39%
Age	Younger <= 24	20%
	Middle = 25 to 64	74%
	Older >= 65	5%

Vehicle

Contributing Factors	Yes	0.1%
	No	96%
	Unknown	4%
Rollover	Yes	2%
	No	98%
Pre-Event Movement	No Driver Present	-
	Going Straight	94%
	Decelerating in Traffic Lane	0.4%
	Accelerating in Traffic Lane	0.1%
	Starting in Traffic Lane	0.1%
	Stopped in Traffic Lane	0.3%
	Passing Another Vehicle	-
	Parked in Travel Lane	-
	Leaving a Parked Position	-
	Entering a Parked Position	-
	Turning Right	-
	Turning Left	-
	Making U-turn	-
	Backing Up	-
	Negotiating a Curve	5%
	Changing Lanes	-
	Merging	-
	Prior Corrective Action	-
	Other	-
Driver Avoidance Maneuver	Object in Road	-
	Poor Road Conditions	-
	Animal in Road	17%
	Vehicle in Road	0.03%
	Non-Motorist in Road	-
	Hit and Run	0.3%
	No Driver Present	-
	Other Avoidance Maneuver	-
	Unknown	69%
	None	13%
	Phantom Vehicle	0.1%
Corrective Action Attempted	No Driver Present	-
	No Avoidance Maneuver	8%
	Braking	4%
	Releasing Brakes	-
	Steering	10%
	Braked and Steered	1%
	Accelerated	-
	Accelerated and Steered	0.01%
	Other	1%
	Unknown	76%

Pedestrian Crash With Prior Vehicle Maneuver

Driving Environment

Category	Value	%
Lighting	Daylight	64%
	Dark Lighted	28%
	Dark	6%
	Dawn/Dusk	2%
Weather	Clear	85%
	Adverse	15%
Road Surface	Dry	78%
	Wet/Slippery	22%
Road Alignment	Straight	96%
	Curve	4%
Road Profile	Level	91%
	Other	9%
Land Use	Rural	27%
	Urban	73%
Day	Weekday	81%
	Weekend	19%
Relation to Roadway	On Roadway	97%
	Shoulder/Parking Lane	1%
	Off Roadway	1%
	Left Turn Lane	1%
	Unknown	-
Relation to Junction	Non-Junction	8%
	Intersection	44%
	Intersection-Related	37%
	Driveway/Alley	9%
	Entrance/Exit Ramp	2%
	Rail Grade Crossing	-
	Other/Unknown	1%
Posted Speed Limit (mph)	<= 20	2%
	25	28%
	30	17%
	35	36%
	40	5%
	45	7%
	50	1%
	>= 55	3%
Traffic Control Device	No Traffic Controls	29%
	Traffic Signal	50%
	Stop/Yield Sign	12%
	Other	9%

Driver

Category	Value	%
Alcohol	Yes	6%
	No	94%
Vision Obscured	No Obstruction	47%
	Vision Obscured	10%
	Unknown	43%
Driver Distracted	Inattention	25%
	Sleepy	-
	Not Distracted	28%
	Unknown	48%
Speeding	Yes	4%
	No	87%
	Unknown	9%
Violation	Speeding	-
	Reckless	0.3%
	None	64%
	Other	26%
	Unknown	10%
Impairment	Ill/Blackout	-
	Drowsy	-
	None	84%
	Other	2%
	Unknown	14%
Gender	Male	62%
	Female	38%
Age	Younger <= 24	16%
	Middle = 25 to 64	72%
	Older >= 65	12%

Vehicle

Category	Value	%
Contributing Factors	Yes	0.2%
	No	84%
	Unknown	16%
Rollover	Yes	0.5%
	No	100%
Pre-Event Movement	No Driver Present	-
	Going Straight	-
	Decelerating in Traffic Lane	-
	Accelerating in Traffic Lane	-
	Starting in Traffic Lane	-
	Stopped in Traffic Lane	-
	Passing Another Vehicle	2%
	Parked in Travel Lane	-
	Leaving a Parked Position	2%
	Entering a Parked Position	0.2%
	Turning Right	33%
	Turning Left	52%
	Making U-turn	0.2%
	Backing Up	1%
	Negotiating a Curve	-
	Changing Lanes	3%
	Merging	-
	Prior Corrective Action	3%
	Other	4%
Driver Avoidance Maneuver	Object in Road	1%
	Poor Road Conditions	-
	Animal in Road	-
	Vehicle in Road	0.3%
	Non-Motorist in Road	10%
	Hit and Run	14%
	No Driver Present	-
	Other Avoidance Maneuver	-
	Unknown	51%
	None	23%
	Phantom Vehicle	-
Corrective Action Attempted	No Driver Present	-
	No Avoidance Maneuver	18%
	Braking	7%
	Releasing Brakes	-
	Steering	4%
	Braked and Steered	0.1%
	Accelerated	-
	Accelerated and Steered	0.1%
	Other	1%
	Unknown	70%

Pedestrian

Location	Intersection – In crosswalk	40%
	Intersection – On roadway	40%
	Intersection – Other	0.1%
	Intersection – Unknown Location	1%
	Non-Intersection – In Crosswalk	0.5%
	Non-Intersection – On Roadway	17%
	Non-Intersection – Other	1%
	Non-Intersection – Unknown Location	0.1%
	In Crosswalk – Unknown if Intersection	-
	Other Location	0.5%
	Unknown Location	0.2%
Action	No Action	69%
	Running Into Road	6%
	Improper Crossing of Roadway	7%
	Inattentive	-
	Jogging	0.2%
	Pushing Vehicle	-
	Walking With Traffic	0.2%
	Walking Against Traffic	0.2%
	Playing in Roadway	9%
	Other Action	1%
	Unknown Action	7%

Pedestrian Crash Without Prior Vehicle Maneuver

Driving Environment

Category	Value	%
Lighting	Daylight	60%
	Dark Lighted	24%
	Dark	9%
	Dawn/Dusk	7%
Weather	Clear	86%
	Adverse	14%
Road Surface	Dry	83%
	Wet/Slippery	17%
Road Alignment	Straight	92%
	Curve	8%
Road Profile	Level	87%
	Other	13%
Land Use	Rural	39%
	Urban	61%
Day	Weekday	79%
	Weekend	21%
Relation to Roadway	On Roadway	96%
	Shoulder/Parking Lane	1%
	Off Roadway	1%
	Left Turn Lane	0.2%
	Unknown	1%
Relation to Junction	Non-Junction	55%
	Intersection	19%
	Intersection-Related	24%
	Driveway/Alley	2%
	Entrance/Exit Ramp	0.3%
	Rail Grade Crossing	-
	Other/Unknown	0.4%
Posted Speed Limit (mph)	<= 20	4%
	25	31%
	30	16%
	35	24%
	40	7%
	45	11%
	50	1%
	>= 55	6%
Traffic Control Device	No Traffic Controls	68%
	Traffic Signal	19%
	Stop/Yield Sign	6%
	Other	7%

Driver

Category	Value	%
Alcohol	Yes	5%
	No	95%
Vision Obscured	No Obstruction	45%
	Vision Obscured	20%
	Unknown	35%
Driver Distracted	Inattention	13%
	Sleepy	0.04%
	Not Distracted	42%
	Unknown	45%
Speeding	Yes	4%
	No	89%
	Unknown	8%
Violation	Speeding	0.2%
	Reckless	1%
	None	74%
	Other	13%
	Unknown	11%
Impairment	Ill/Blackout	-
	Drowsy	-
	None	90%
	Other	2%
	Unknown	8%
Gender	Male	60%
	Female	40%
Age	Younger <= 24	29%
	Middle = 25 to 64	60%
	Older >= 65	11%

Vehicle

Category	Value	%
Contributing Factors	Yes	0.1%
	No	84%
	Unknown	16%
Rollover	Yes	0.1%
	No	100%
Pre-Event Movement	No Driver Present	-
	Going Straight	92%
	Decelerating in Traffic Lane	1%
	Accelerating in Traffic Lane	0.2%
	Starting in Traffic Lane	3%
	Stopped in Traffic Lane	1%
	Passing Another Vehicle	-
	Parked in Travel Lane	-
	Leaving a Parked Position	-
	Entering a Parked Position	-
	Turning Right	-
	Turning Left	-
	Making U-turn	-
	Backing Up	-
	Negotiating a Curve	3%
	Changing Lanes	-
	Merging	-
	Prior Corrective Action	-
	Other	-
Driver Avoidance Maneuver	Object in Road	-
	Poor Road Conditions	-
	Animal in Road	0.1%
	Vehicle in Road	1%
	Non-Motorist in Road	22%
	Hit and Run	13%
	No Driver Present	-
	Other Avoidance Maneuver	0.1%
	Unknown	47%
	None	17%
	Phantom Vehicle	-
Corrective Action Attempted	No Driver Present	-
	No Avoidance Maneuver	15%
	Braking	14%
	Releasing Brakes	-
	Steering	6%
	Braked and Steered	4%
	Accelerated	0.04%
	Accelerated and Steered	0.1%
	Other	1%
	Unknown	61%

Pedestrian

Location	Intersection – In Crosswalk	13%
	Intersection – On Roadway	25%
	Intersection – Other	1%
	Intersection – Unknown Location	1%
	Non-Intersection – In Crosswalk	1%
	Non-Intersection – On Roadway	57%
	Non-Intersection – Other	1%
	Non-Intersection – Unknown Location	0.1%
	In Crosswalk – Unknown if Intersection	-
	Other Location	1%
	Unknown Location	1%
Action	No Action	17%
	Running Into Road	36%
	Improper Crossing of Roadway	26%
	Inattentive	1.3%
	Jogging	0.1%
	Pushing Vehicle	0.04%
	Walking With Traffic	2.4%
	Walking Against Traffic	1.0%
	Playing in Roadway	8%
	Other Action	6%
	Unknown Action	2%

Pedalcyclist Crash With Prior Vehicle Maneuver

Driving Environment

Lighting	Daylight	78%
	Dark Lighted	14%
	Dark	2%
	Dawn/Dusk	6%
Weather	Clear	97%
	Adverse	3%
Road Surface	Dry	93%
	Wet/Slippery	7%
Road Alignment	Straight	91%
	Curve	9%
Road Profile	Level	83%
	Other	17%
Land Use	Rural	47%
	Urban	53%
Day	Weekday	81%
	Weekend	19%
Relation to Roadway	On Roadway	97%
	Shoulder/Parking Lane	1%
	Off Roadway	1%
	Left Turn Lane	0.2%
	Unknown	0.1%
Relation to Junction	Non-Junction	2%
	Intersection	47%
	Intersection-Related	30%
	Driveway/Alley	19%
	Entrance/Exit Ramp	0.2%
	Rail Grade Crossing	-
	Other/Unknown	2%
Posted Speed Limit (mph)	<= 20	8%
	25	30%
	30	17%
	35	28%
	40	6%
	45	8%
	50	2%
	>= 55	2%
Traffic Control Device	No Traffic Controls	33%
	Traffic Signal	34%
	Stop/Yield Sign	26%
	Other	8%

Driver

Alcohol	Yes	3%
	No	97%
Vision Obscured	No Obstruction	45%
	Vision Obscured	11%
	Unknown	44%
Driver Distracted	Inattention	25%
	Sleepy	-
	Not Distracted	35%
	Unknown	40%
Speeding	Yes	0.1%
	No	89%
	Unknown	11%
Violation	Speeding	-
	Reckless	-
	None	60%
	Other	26%
	Unknown	13%
Impairment	Ill/Blackout	-
	Drowsy	-
	None	91%
	Other	1%
	Unknown	7%
Gender	Male	61%
	Female	39%
Age	Younger <= 24	28%
	Middle = 25 to 64	58%
	Older >= 65	14%

Vehicle

Contributing Factors	Yes	0.1%
	No	82%
	Unknown	18%
Rollover	Yes	-
	No	100%
Pre-Event Movement	No Driver Present	-
	Going Straight	-
	Decelerating in Traffic Lane	-
	Accelerating in Traffic Lane	-
	Starting in Traffic Lane	-
	Stopped in Traffic Lane	-
	Passing Another Vehicle	3%
	Parked in Travel Lane	-
	Leaving a Parked Position	2%
	Entering a Parked Position	0.1%
	Turning Right	55%
	Turning Left	34%
	Making U-turn	1%
	Backing Up	-
	Negotiating a Curve	-
	Changing Lanes	0.4%
	Merging	0.2%
	Prior Corrective Action	0.1%
	Other	5%
Driver Avoidance Maneuver	Object in Road	-
	Poor Road Conditions	-
	Animal in Road	-
	Vehicle in Road	0.1%
	Non-Motorist in Road	5%
	Hit and Run	15%
	No Driver Present	-
	Other Avoidance Maneuver	-
	Unknown	56%
	None	24%
	Phantom Vehicle	-
Corrective Action Attempted	No Driver Present	-
	No Avoidance Maneuver	23%
	Braking	4%
	Releasing Brakes	-
	Steering	0.3%
	Braked and Steered	0.1%
	Accelerated	0.4%
	Accelerated and Steered	-
	Other	0.2%
	Unknown	72%

Pedalcyclist

Location	Intersection – In Crosswalk	31%
	Intersection – On Roadway	43%
	Intersection – Other	2%
	Intersection – Unknown Location	1%
	Non-Intersection – In Crosswalk	1%
	Non-Intersection – On Roadway	19%
	Non-Intersection – Other	0.3%
	Non-Intersection – Unknown Location	0.4%
	In Crosswalk – Unknown If Intersection	-
	Other Location	1%
	Unknown Location	1%
Action	No Action	48%
	Failing to Have Lights On	1%
	Operating Without Required Equipment	1%
	Improper Lane Changing	-
	Failure to Keep in Proper Lane or Road	0.4%
	Making Improper Entry/Exit	0.2%
	Operating the Vehicle in Reckless Manner	1%
	Failure to Yield Right-of-Way	13%
	Failure to Obey Traffic Signs	2%
	Making Other Improper Turn	0.1%
	Driving on Wrong Side of Road	24%
	Other Action	6%
	Unknown Action	4%

Pedalcyclist Crash Without Prior Vehicle Maneuver

Driving Environment

Lighting	Daylight	73%
	Dark Lighted	17%
	Dark	5%
	Dawn/Dusk	5%
Weather	Clear	93%
	Adverse	7%
Road Surface	Dry	90%
	Wet/Slippery	10%
Road Alignment	Straight	91%
	Curve	9%
Road Profile	Level	80%
	Other	20%
Land Use	Rural	45%
	Urban	55%
Day	Weekday	72%
	Weekend	28%
Relation to Roadway	On Roadway	97%
	Shoulder/Parking Lane	1%
	Off Roadway	2%
	Left Turn Lane	-
	Unknown	0.4%
Relation to Junction	Non-Junction	31%
	Intersection	40%
	Intersection-Related	15%
	Driveway/Alley	13%
	Entrance/Exit Ramp	0.1%
	Rail Grade Crossing	-
	Other/Unknown	2%
Posted Speed Limit (mph)	<= 20	6%
	25	32%
	30	13%
	35	22%
	40	5%
	45	14%
	50	0.4%
	>= 55	7%
Traffic Control Device	No Traffic Controls	53%
	Traffic Signal	17%
	Stop/Yield Sign	26%
	Other	4%

Driver

Alcohol	Yes	4%
	No	96%
Vision Obscured	No Obstruction	55%
	Vision Obscured	17%
	Unknown	29%
Driver Distracted	Inattention	14%
	Sleepy	0.2%
	Not Distracted	47%
	Unknown	39%
Speeding	Yes	2%
	No	91%
	Unknown	8%
Violation	Speeding	-
	Reckless	0.3%
	None	75%
	Other	17%
	Unknown	8%
Impairment	Ill/Blackout	-
	Drowsy	-
	None	93%
	Other	1%
	Unknown	6%
Gender	Male	50%
	Female	50%
Age	Younger <= 24	20%
	Middle = 25 to 64	69%
	Older >= 65	11%

Vehicle

Contributing Factors	Yes	0.1%
	No	91%
	Unknown	9%
Rollover	Yes	0.2%
	No	100%
Pre-Event Movement	No Driver Present	-
	Going Straight	80%
	Decelerating in Traffic Lane	1%
	Accelerating in Traffic Lane	0.1%
	Starting in Traffic Lane	9%
	Stopped in Traffic Lane	6%
	Passing Another Vehicle	-
	Parked in Travel Lane	-
	Leaving a Parked Position	-
	Entering a Parked Position	-
	Turning Right	-
	Turning Left	-
	Making U-turn	-
	Backing Up	-
	Negotiating a Curve	3%
	Changing Lanes	-
	Merging	-
	Prior Corrective Action	-
	Other	-
Driver Avoidance Maneuver	Object in Road	-
	Poor Road Conditions	-
	Animal in Road	-
	Vehicle in Road	1%
	Non-Motorist in Road	20%
	Hit and Run	8%
	No Driver Present	-
	Other Avoidance Maneuver	-
	Unknown	47%
	None	24%
	Phantom Vehicle	0.1%
Corrective Action Attempted	No Driver Present	-
	No Avoidance Maneuver	23%
	Braking	9%
	Releasing Brakes	-
	Steering	7%
	Braked and Steered	4%
	Accelerated	0.1%
	Accelerated and Steered	0.1%
	Other	0.3%
	Unknown	56%

Pedalcyclist

Location	Intersection – In Crosswalk	6%
	Intersection – On Roadway	47%
	Intersection – Other	1%
	Intersection – Unknown Location	0.1%
	Non-Intersection – In Crosswalk	2%
	Non-intersection – On Roadway	43%
	Non-Intersection – Other	0.2%
	Non-Intersection – Unknown Location	0.2%
	In Crosswalk – Unknown If Intersection	0.1%
	Other Location	0.1%
	Unknown Location	0.5%
Action	No Action	19%
	Failing to Have Lights On	3%
	Operating Without Required Equipment	2%
	Improper Lane Changing	2%
	Failure to Keep in Proper Lane or Road	1%
	Making Improper Entry/Exit	4%
	Operating the Vehicle in Reckless Manner	2%
	Failure to Yield Right-of-Way	46%
	Failure to Obey Traffic Signs	1%
	Making Other Improper Turn	1%
	Driving on Wrong Side of Road	6%
	Other Action	7%
	Unknown Action	5%

Backing Up Into Another Vehicle

Driving Environment

Category	Value	%
Lighting	Daylight	83%
	Dark Lighted	11%
	Dark	4%
	Dawn/Dusk	2%
Weather	Clear	89%
	Adverse	11%
Road Surface	Dry	83%
	Wet/Slippery	17%
Road Alignment	Straight	93%
	Curve	7%
Road Profile	Level	81%
	Other	19%
Land Use	Rural	48%
	Urban	52%
Day	Weekday	80%
	Weekend	20%
Relation to Roadway	On Roadway	98%
	Shoulder/Parking Lane	1%
	Off Roadway	0.4%
	Left Turn Lane	0.2%
	Unknown	0.1%
Relation to Junction	Non-Junction	25%
	Intersection	5%
	Intersection-Related	27%
	Driveway/Alley	38%
	Entrance/Exit Ramp	2%
	Rail Grade Crossing	1%
	Other/Unknown	3%
Posted Speed Limit (mph)	<= 20	8%
	25	38%
	30	13%
	35	19%
	40	5%
	45	7%
	50	1%
	>= 55	8%
Traffic Control Device	No Traffic Controls	66%
	Traffic Signal	16%
	Stop/Yield Sign	11%
	Other	6%

Driver

Category	Value	%
Alcohol	Yes	2%
	No	98%
Vision Obscured	No Obstruction	59%
	Vision Obscured	10%
	Unknown	31%
Driver Distracted	Inattention	34%
	Sleepy	0.2%
	Not Distracted	22%
	Unknown	43%
Speeding	Yes	1%
	No	96%
	Unknown	4%
Violation	Speeding	-
	Reckless	0.1%
	None	54%
	Other	40%
	Unknown	6%
Impairment	Ill/Blackout	-
	Drowsy	0.2%
	None	94%
	Other	2%
	Unknown	4%
Gender	Male	62%
	Female	38%
Age	Younger <= 24	24%
	Middle = 25 to 64	66%
	Older >= 65	10%

Vehicle

Category	Value	%
Contributing Factors	Yes	0.4%
	No	90%
	Unknown	9%
Rollover	Yes	-
	No	100%
Pre-Event Movement	No Driver Present	0.2%
	Going Straight	0.02%
	Decelerating in Traffic Lane	-
	Accelerating in Traffic Lane	-
	Starting in Traffic Lane	0.2%
	Stopped in Traffic Lane	6%
	Passing Another Vehicle	-
	Parked in Travel Lane	-
	Leaving a Parked Position	11%
	Entering a Parked Position	2%
	Turning Right	-
	Turning Left	-
	Making U-turn	-
	Backing Up	79%
	Negotiating a Curve	-
	Changing Lanes	-
	Merging	-
	Prior Corrective Action	-
	Other	2%
Driver Avoidance Maneuver	Object in Road	-
	Poor Road Conditions	-
	Animal in Road	-
	Vehicle in Road	1%
	Non-Motorist in Road	-
	Hit and Run	7%
	No Driver Present	0.2%
	Other Avoidance Maneuver	0.004%
	Unknown	56%
	None	35%
	Phantom Vehicle	0.2%
Corrective Action Attempted	No Driver Present	0.2%
	No Avoidance Maneuver	31%
	Braking	0.5%
	Releasing Brakes	-
	Steering	-
	Braked and Steered	-
	Accelerated	-
	Accelerated and Steered	-
	Other	1%
	Unknown	67%

Driver and vehicle statistics represent the backing light vehicle.

Vehicle(s) Turning – Vehicles Traveling in Same Direction

Driving Environment

Lighting	Daylight	79%
	Dark Lighted	14%
	Dark	4%
	Dawn/Dusk	3%
Weather	Clear	90%
	Adverse	10%
Road Surface	Dry	84%
	Wet/Slippery	16%
Road Alignment	Straight	92%
	Curve	8%
Road Profile	Level	83%
	Other	17%
Land Use	Rural	49%
	Urban	51%
Day	Weekday	79%
	Weekend	21%
Relation to Roadway	On Roadway	99%
	Shoulder/Parking Lane	1%
	Off Roadway	0.2%
	Left Turn Lane	1%
	Unknown	0.1%
Relation to Junction	Non-Junction	6%
	Intersection	37%
	Intersection-Related	22%
	Driveway/Alley	28%
	Entrance/Exit Ramp	1%
	Rail Grade Crossing	-
	Other/Unknown	5%
Posted Speed Limit (mph)	<= 20	2%
	25	17%
	30	16%
	35	27%
	40	9%
	45	16%
	50	3%
	>= 55	11%
Traffic Control Device	No Traffic Controls	60%
	Traffic Signal	29%
	Stop/Yield Sign	6%
	Other	5%

Driver

Alcohol	Yes	2%
	No	98%
Vision Obscured	No Obstruction	73%
	Vision Obscured	1%
	Unknown	26%
Driver Distracted	Inattention	14%
	Sleepy	0.1%
	Not Distracted	42%
	Unknown	44%
Speeding	Yes	1%
	No	96%
	Unknown	3%
Violation	Speeding	-
	Reckless	0.4%
	None	69%
	Other	28%
	Unknown	3%
Impairment	Ill/Blackout	-
	Drowsy	0.1%
	None	96%
	Other	1%
	Unknown	3%
Gender	Male	59%
	Female	41%
Age	Younger <= 24	26%
	Middle = 25 to 64	63%
	Older >= 65	11%

Vehicle

Contributing Factors	Yes	1%
	No	93%
	Unknown	6%
Rollover	Yes	0.2%
	No	100%
Pre-Event Movement	No Driver Present	-
	Going Straight	-
	Decelerating in Traffic Lane	-
	Accelerating in Traffic Lane	-
	Starting in Traffic Lane	-
	Stopped in Traffic Lane	-
	Passing Another Vehicle	-
	Parked in Travel Lane	-
	Leaving a Parked Position	-
	Entering a Parked Position	-
	Turning Right	40%
	Turning Left	52%
	Making U-turn	8%
	Backing Up	-
	Negotiating a Curve	-
	Changing Lanes	-
	Merging	-
	Prior Corrective Action	-
	Other	-
Driver Avoidance Maneuver	Object in Road	-
	Poor Road Conditions	-
	Animal in Road	-
	Vehicle in Road	1%
	Non-Motorist in Road	-
	Hit and Run	4%
	No Driver Present	-
	Other Avoidance Maneuver	-
	Unknown	70%
	None	25%
	Phantom Vehicle	-
Corrective Action Attempted	No Driver Present	-
	No Avoidance Maneuver	21%
	Braking	0.5%
	Releasing Brakes	-
	Steering	0.2%
	Braked and Steered	0.03%
	Accelerated	0.1%
	Accelerated and Steered	-
	Other	-
	Unknown	78%

Driver and vehicle statistics represent the turning light vehicle.

Vehicle(s) Parking – Vehicles Traveling in Same Direction

Driving Environment

Lighting	Daylight	82%
	Dark Lighted	12%
	Dark	5%
	Dawn/Dusk	1%
Weather	Clear	85%
	Adverse	15%
Road Surface	Dry	76%
	Wet/Slippery	24%
Road Alignment	Straight	91%
	Curve	9%
Road Profile	Level	85%
	Other	15%
Land Use	Rural	39%
	Urban	61%
Day	Weekday	84%
	Weekend	16%
Relation to Roadway	On Roadway	97%
	Shoulder/Parking Lane	1%
	Off Roadway	1%
	Left Turn Lane	0.2%
	Unknown	0.5%
Relation to Junction	Non-Junction	74%
	Intersection	3%
	Intersection-Related	10%
	Driveway/Alley	2%
	Entrance/Exit Ramp	5%
	Rail Grade Crossing	-
	Other/Unknown	5%
Posted Speed Limit (mph)	<= 20	2%
	25	27%
	30	12%
	35	21%
	40	6%
	45	10%
	50	1%
	>= 55	20%
Traffic Control Device	No Traffic Controls	81%
	Traffic Signal	9%
	Stop/Yield Sign	3%
	Other	7%

Driver

Alcohol	Yes	2%
	No	98%
Vision Obscured	No Obstruction	57%
	Vision Obscured	4%
	Unknown	39%
Driver Distracted	Inattention	22%
	Sleepy	-
	Not Distracted	37%
	Unknown	41%
Speeding	Yes	3%
	No	93%
	Unknown	4%
Violation	Speeding	-
	Reckless	0.3%
	None	65%
	Other	30%
	Unknown	4%
Impairment	Ill/Blackout	-
	Drowsy	-
	None	96%
	Other	1%
	Unknown	3%
Gender	Male	59%
	Female	41%
Age	Younger <= 24	27%
	Middle = 25 to 64	63%
	Older >= 65	10%

Vehicle

Contributing Factors	Yes	0.01%
	No	94%
	Unknown	6%
Rollover	Yes	0.3%
	No	100%
Pre-Event Movement	No Driver Present	0.3%
	Going Straight	3%
	Decelerating in Traffic Lane	-
	Accelerating in Traffic Lane	-
	Starting in Traffic Lane	-
	Stopped in Traffic Lane	0.04%
	Passing Another Vehicle	2%
	Parked in Travel Lane	-
	Leaving a Parked Position	68%
	Entering a Parked Position	8%
	Turning Right	1%
	Turning Left	0.5%
	Making U-turn	10%
	Backing Up	-
	Negotiating a Curve	2%
	Changing Lanes	0.02%
	Merging	5%
	Prior Corrective Action	-
	Other	1%
Driver Avoidance Maneuver	Object in Road	-
	Poor Road Conditions	-
	Animal in Road	-
	Vehicle in Road	4%
	Non-Motorist in Road	-
	Hit and Run	5%
	No Driver Present	0.3%
	Other Avoidance Maneuver	-
	Unknown	76%
	None	14%
	Phantom Vehicle	-
Corrective Action Attempted	No Driver Present	0.3%
	No Avoidance Maneuver	9%
	Braking	1%
	Releasing Brakes	-
	Steering	2%
	Braked and Steered	1%
	Accelerated	0.2%
	Accelerated and Steered	-
	Other	0.04%
	Unknown	86%

Driver and vehicle statistics represent the parking light vehicle.

Vehicle(s) Changing Lanes – Vehicles Traveling in Same Direction

Driving Environment

Lighting	Daylight	74%
	Dark Lighted	17%
	Dark	6%
	Dawn/Dusk	3%
Weather	Clear	89%
	Adverse	11%
Road Surface	Dry	83%
	Wet/Slippery	17%
Road Alignment	Straight	90%
	Curve	10%
Road Profile	Level	81%
	Other	19%
Land Use	Rural	46%
	Urban	54%
Day	Weekday	78%
	Weekend	22%
Relation to Roadway	On Roadway	90%
	Shoulder/Parking Lane	1%
	Off Roadway	8%
	Left Turn Lane	1%
	Unknown	0.1%
Relation to Junction	Non-Junction	69%
	Intersection	4%
	Intersection-Related	15%
	Driveway/Alley	1%
	Entrance/Exit Ramp	9%
	Rail Grade Crossing	0.1%
	Other/Unknown	2%
Posted Speed Limit (mph)	<= 20	1%
	25	7%
	30	6%
	35	20%
	40	10%
	45	17%
	50	4%
	>= 55	34%
Traffic Control Device	No Traffic Controls	79%
	Traffic Signal	11%
	Stop/Yield Sign	2%
	Other	8%

Driver

Alcohol	Yes	3%
	No	97%
Vision Obscured	No Obstruction	65%
	Vision Obscured	2%
	Unknown	33%
Driver Distracted	Inattention	22%
	Sleepy	0.1%
	Not Distracted	29%
	Unknown	49%
Speeding	Yes	4%
	No	88%
	Unknown	7%
Violation	Speeding	-
	Reckless	1%
	None	53%
	Other	37%
	Unknown	9%
Impairment	Ill/Blackout	0.1%
	Drowsy	0.1%
	None	92%
	Other	1%
	Unknown	6%
Gender	Male	59%
	Female	41%
Age	Younger <= 24	32%
	Middle = 25 to 64	58%
	Older >= 65	10%

Vehicle

Contributing Factors	Yes	1%
	No	86%
	Unknown	14%
Rollover	Yes	2%
	No	98%
Pre-Event Movement	No Driver Present	-
	Going Straight	6%
	Decelerating in Traffic Lane	0.002%
	Accelerating in Traffic Lane	-
	Starting in Traffic Lane	-
	Stopped in Traffic Lane	-
	Passing Another Vehicle	15%
	Parked in Travel Lane	-
	Leaving a Parked Position	-
	Entering a Parked Position	-
	Turning Right	-
	Turning Left	0.1%
	Making U-turn	-
	Backing Up	-
	Negotiating a Curve	1%
	Changing Lanes	69%
	Merging	8%
	Prior Corrective Action	1%
	Other	-
Driver Avoidance Maneuver	Object in Road	0.1%
	Poor Road Conditions	0.01%
	Animal in Road	-
	Vehicle in Road	11%
	Non-Motorist in Road	-
	Hit and Run	11%
	No Driver Present	-
	Other Avoidance Maneuver	-
	Unknown	55%
	None	22%
	Phantom Vehicle	2%
Corrective Action Attempted	No Driver Present	-
	No Avoidance Maneuver	19%
	Braking	1%
	Releasing Brakes	-
	Steering	11%
	Braked and Steered	1%
	Accelerated	0.001%
	Accelerated and Steered	0.002%
	Other	0.02%
	Unknown	68%

Driver and vehicle statistics represent the light vehicle changing lanes.

Vehicle(s) Drifting – Vehicles Traveling in Same Direction

Driving Environment

Category	Value	%
Lighting	Daylight	74%
	Dark Lighted	18%
	Dark	5%
	Dawn/Dusk	4%
Weather	Clear	84%
	Adverse	16%
Road Surface	Dry	78%
	Wet/Slippery	22%
Road Alignment	Straight	87%
	Curve	13%
Road Profile	Level	80%
	Other	20%
Land Use	Rural	45%
	Urban	55%
Day	Weekday	80%
	Weekend	20%
Relation to Roadway	On Roadway	99%
	Shoulder/Parking Lane	0.01%
	Off Roadway	0.3%
	Left Turn Lane	1%
	Unknown	0.2%
Relation to Junction	Non-Junction	64%
	Intersection	5%
	Intersection-Related	21%
	Driveway/Alley	2%
	Entrance/Exit Ramp	5%
	Rail Grade Crossing	1%
	Other/Unknown	3%
Posted Speed Limit (mph)	<= 20	1%
	25	5%
	30	8%
	35	22%
	40	9%
	45	17%
	50	4%
	>= 55	33%
Traffic Control Device	No Traffic Controls	73%
	Traffic Signal	16%
	Stop/Yield Sign	3%
	Other	8%

Driver

Category	Value	%
Alcohol	Yes	4%
	No	96%
Vision Obscured	No Obstruction	71%
	Vision Obscured	1%
	Unknown	28%
Driver Distracted	Inattention	10%
	Sleepy	1%
	Not Distracted	43%
	Unknown	47%
Speeding	Yes	8%
	No	84%
	Unknown	8%
Violation	Speeding	-
	Reckless	1%
	None	70%
	Other	21%
	Unknown	9%
Impairment	Ill/Blackout	0.4%
	Drowsy	1%
	None	91%
	Other	2%
	Unknown	6%
Gender	Male	60%
	Female	40%
Age	Younger <= 24	27%
	Middle = 25 to 64	64%
	Older >= 65	9%

Vehicle

Category	Value	%
Contributing Factors	Yes	1%
	No	88%
	Unknown	11%
Rollover	Yes	1%
	No	99%
Pre-Event Movement	No Driver Present	2%
	Going Straight	68%
	Decelerating in Traffic Lane	7%
	Accelerating in Traffic Lane	0.1%
	Starting in Traffic Lane	1%
	Stopped in Traffic Lane	12%
	Passing Another Vehicle	-
	Parked in Travel Lane	1%
	Leaving a Parked Position	-
	Entering a Parked Position	-
	Turning Right	-
	Turning Left	-
	Making U-turn	-
	Backing Up	-
	Negotiating a Curve	5%
	Changing Lanes	-
	Merging	-
	Prior Corrective Action	2%
	Other	3%
Driver Avoidance Maneuver	Object in Road	1%
	Poor Road Conditions	0.01%
	Animal in Road	-
	Vehicle in Road	19%
	Non-Motorist in Road	-
	Hit and Run	8%
	No Driver Present	2%
	Other Avoidance Maneuver	0.01%
	Unknown	47%
	None	24%
	Phantom Vehicle	0.2%
Corrective Action Attempted	No Driver Present	2%
	No Avoidance Maneuver	19%
	Braking	3%
	Releasing Brakes	-
	Steering	16%
	Braked and Steered	2%
	Accelerated	0.03%
	Accelerated and Steered	0.003%
	Other	0.3%
	Unknown	59%

Driver and vehicle statistics represent all light vehicles involved.

Vehicle(s) Making a Maneuver – Vehicles Traveling in Opposite Direction

Driving Environment			*Driver*			*Vehicle*		
Lighting	Daylight	60%	Alcohol	Yes	16%	Contributing Factors	Yes	3%
	Dark Lighted	18%		No	84%		No	81%
	Dark	21%	Vision Obscured	No Obstruction	59%		Unknown	16%
	Dawn/Dusk	1%		Vision Obscured	7%	Rollover	Yes	4%
Weather	Clear	82%		Unknown	34%		No	96%
	Adverse	18%	Driver Distracted	Inattention	18%	Pre-Event Movement	No Driver Present	-
Road Surface	Dry	73%		Sleepy	2%		Going Straight	-
	Wet/Slippery	27%		Not Distracted	35%		Decelerating in Traffic Lane	-
Road Alignment	Straight	87%		Unknown	46%		Accelerating in Traffic Lane	-
	Curve	13%	Speeding	Yes	10%		Starting in Traffic Lane	-
Road Profile	Level	70%		No	83%		Stopped in Traffic Lane	-
	Other	30%		Unknown	7%		Passing Another Vehicle	34%
Land Use	Rural	53%	Violation	Speeding	0.3%		Parked in Travel Lane	-
	Urban	47%		Reckless	1%		Leaving a Parked Position	6%
Day	Weekday	69%		None	51%		Entering a Parked Position	2%
	Weekend	31%		Other	38%		Turning Right	-
Relation to Roadway	On Roadway	78%		Unknown	9%		Turning Left	1%
	Shoulder/Parking Lane	3%	Impairment	Ill/Blackout	1%		Making U-turn	-
	Off Roadway	14%		Drowsy	2%		Backing Up	-
	Left Turn Lane	5%		None	87%		Negotiating a Curve	-
	Unknown	-		Other	9%		Changing Lanes	12%
Relation to Junction	Non-Junction	81%		Unknown	2%		Merging	2%
	Intersection	7%	Gender	Male	72%		Prior Corrective Action	16%
	Intersection-Related	10%		Female	28%		Other	28%
	Driveway/Alley	1%	Age	Younger <= 24	65%	Driver Avoidance Maneuver	Object in Road	-
	Entrance/Exit Ramp	0.1%		Middle = 25 to 64	29%		Poor Road Conditions	-
	Rail Grade Crossing	-		Older >= 65	6%		Animal in Road	-
	Other/Unknown	1%					Vehicle in Road	26%
Posted Speed Limit (mph)	<= 20	1%					Non-Motorist in Road	-
	25	11%					Hit and Run	13%
	30	9%					No Driver Present	-
	35	18%					Other Avoidance Maneuver	-
	40	6%					Unknown	45%
	45	20%					None	13%
	50	6%					Phantom Vehicle	2%
	>= 55	29%				Corrective Action Attempted	No Driver Present	-
Traffic Control Device	No Traffic Controls	86%					No Avoidance Maneuver	9%
	Traffic Signal	9%					Braking	3%
	Stop/Yield Sign	0.03%					Releasing Brakes	-
	Other	5%					Steering	24%
							Braked and Steered	1%
							Accelerated	-
							Accelerated and Steered	0.04%
							Other	1%
							Unknown	63%

Driver and vehicle statistics represent the light vehicle making a maneuver.

Vehicle(s) Not Making a Maneuver – Vehicles Traveling in Opposite Direction

Driving Environment   Driver   Vehicle

Category	Value	%	Category	Value	%	Category	Value	%
Lighting	Daylight	65%	Alcohol	Yes	6%	Contributing Factors	Yes	1%
	Dark Lighted	11%		No	94%		No	90%
	Dark	18%	Vision Obscured	No Obstruction	68%		Unknown	9%
	Dawn/Dusk	6%		Vision Obscured	5%	Rollover	Yes	3%
Weather	Clear	78%		Unknown	28%		No	97%
	Adverse	22%	Driver Distracted	Inattention	8%	Pre-Event Movement	No Driver Present	1%
Road Surface	Dry	70%		Sleepy	2%		Going Straight	63%
	Wet/Slippery	30%		Not Distracted	44%		Decelerating in Traffic Lane	1%
Road Alignment	Straight	58%		Unknown	46%		Accelerating in Traffic Lane	0.01%
	Curve	42%	Speeding	Yes	7%		Starting in Traffic Lane	0.2%
Road Profile	Level	65%		No	88%		Stopped in Traffic Lane	3%
	Other	35%		Unknown	5%		Passing Another Vehicle	-
Land Use	Rural	66%	Violation	Speeding	0.2%		Parked in Travel Lane	0.2%
	Urban	34%		Reckless	1%		Leaving a Parked Position	-
Day	Weekday	71%		None	71%		Entering a Parked Position	-
	Weekend	29%		Other	22%		Turning Right	-
Relation to Roadway	On Roadway	81%		Unknown	6%		Turning Left	-
	Shoulder/Parking Lane	2%	Impairment	Ill/Blackout	0.4%		Making U-turn	-
	Off Roadway	17%		Drowsy	2%		Backing Up	-
	Left Turn Lane	0.3%		None	88%		Negotiating a Curve	32%
	Unknown	0.2%		Other	4%		Changing Lanes	-
Relation to Junction	Non-Junction	88%		Unknown	6%		Merging	-
	Intersection	3%	Gender	Male	63%		Prior Corrective Action	-
	Intersection-Related	7%		Female	37%		Other	-
	Driveway/Alley	0.4%	Age	Younger <= 24	29%	Driver Avoidance Maneuver	Object in Road	0.2%
	Entrance/Exit Ramp	0.1%		Middle = 25 to 64	64%		Poor Road Conditions	0.001%
	Rail Grade Crossing	0.2%		Older >= 65	7%		Animal in Road	-
	Other/Unknown	1%					Vehicle in Road	21%
Posted Speed Limit (mph)	<= 20	3%					Non-Motorist in Road	0.1%
	25	17%					Hit and Run	6%
	30	11%					No Driver Present	1%
	35	21%					Other Avoidance Maneuver	0.1%
	40	5%					Unknown	56%
	45	13%					None	14%
	50	3%					Phantom Vehicle	2%
	>= 55	28%				Corrective Action Attempted	No Driver Present	1%
Traffic Control Device	No Traffic Controls	86%					No Avoidance Maneuver	10%
	Traffic Signal	4%					Braking	5%
	Stop/Yield Sign	1%					Releasing Brakes	-
	Other	9%					Steering	18%
							Braked and Steered	2%
							Accelerated	-
							Accelerated and Steered	0.1%
							Other	0.3%
							Unknown	64%

Driver and vehicle statistics represent all light vehicles involved.

Following Vehicle Making a Maneuver and Approaching Lead Vehicle

Driving Environment

Lighting	Daylight	76%
	Dark Lighted	16%
	Dark	3%
	Dawn/Dusk	4%
Weather	Clear	91%
	Adverse	9%
Road Surface	Dry	85%
	Wet/Slippery	15%
Road Alignment	Straight	84%
	Curve	16%
Road Profile	Level	80%
	Other	20%
Land Use	Rural	42%
	Urban	58%
Day	Weekday	77%
	Weekend	23%
Relation to Roadway	On Roadway	96%
	Shoulder/Parking Lane	1%
	Off Roadway	2%
	Left Turn Lane	0.3%
	Unknown	-
Relation to Junction	Non-Junction	36%
	Intersection	7%
	Intersection-Related	33%
	Driveway/Alley	4%
	Entrance/Exit Ramp	6%
	Rail Grade Crossing	0.3%
	Other/Unknown	14%
Posted Speed Limit (mph)	<= 20	0.5%
	25	8%
	30	9%
	35	25%
	40	10%
	45	19%
	50	5%
	>= 55	24%
Traffic Control Device	No Traffic Controls	50%
	Traffic Signal	29%
	Stop/Yield Sign	14%
	Other	7%

Driver

Alcohol	Yes	5%
	No	95%
Vision Obscured	No Obstruction	64%
	Vision Obscured	2%
	Unknown	34%
Driver Distracted	Inattention	29%
	Sleepy	0.3%
	Not Distracted	24%
	Unknown	47%
Speeding	Yes	25%
	No	64%
	Unknown	11%
Violation	Speeding	-
	Reckless	1%
	None	44%
	Other	42%
	Unknown	13%
Impairment	Ill/Blackout	0.1%
	Drowsy	0.2%
	None	88%
	Other	2%
	Unknown	10%
Gender	Male	59%
	Female	41%
Age	Younger <= 24	33%
	Middle = 25 to 64	62%
	Older >= 65	5%

Vehicle

Contributing Factors	Yes	1%
	No	80%
	Unknown	20%
Rollover	Yes	0.1%
	No	100%
Pre-Event Movement	No Driver Present	-
	Going Straight	-
	Decelerating in Traffic Lane	-
	Accelerating in Traffic Lane	-
	Starting in Traffic Lane	-
	Stopped in Traffic Lane	-
	Passing Another Vehicle	9%
	Parked in Travel Lane	-
	Leaving a Parked Position	6%
	Entering a Parked Position	1%
	Turning Right	22%
	Turning Left	7%
	Making U-turn	0.3%
	Backing Up	-
	Negotiating a Curve	-
	Changing Lanes	36%
	Merging	4%
	Prior Corrective Action	3%
	Other	12%
Driver Avoidance Maneuver	Object in Road	0.2%
	Poor Road Conditions	-
	Animal in Road	-
	Vehicle in Road	12%
	Non-Motorist in Road	-
	Hit and Run	17%
	No Driver Present	-
	Other Avoidance Maneuver	-
	Unknown	57%
	None	13%
	Phantom Vehicle	0.01%
Corrective Action Attempted	No Driver Present	-
	No Avoidance Maneuver	11%
	Braking	5%
	Releasing Brakes	-
	Steering	8%
	Braked and Steered	1%
	Accelerated	0.2%
	Accelerated and Steered	-
	Other	0.1%
	Unknown	75%

Driver and vehicle statistics represent the striking light vehicle.

Following Vehicle Approaching an Accelerating Lead Vehicle

Driving Environment *Driver* *Vehicle*

Lighting	Daylight	78%	**Alcohol**	Yes	3%	**Contributing Factors**	Yes	1%
	Dark Lighted	12%		No	97%		No	88%
	Dark	5%	**Vision Obscured**	No Obstruction	71%		Unknown	11%
	Dawn/Dusk	5%		Vision Obscured	1%	**Rollover**	Yes	0.01%
Weather	Clear	91%		Unknown	28%		No	100%
	Adverse	9%	**Driver Distracted**	Inattention	39%	**Pre-Event Movement**	No Driver Present	-
Road Surface	Dry	89%		Sleepy	2%		Going Straight	54%
	Wet/Slippery	11%		Not Distracted	21%		Decelerating in Traffic Lane	5%
Road Alignment	Straight	91%		Unknown	38%		Accelerating in Traffic Lane	3%
	Curve	9%	**Speeding**	Yes	30%		Starting in Traffic Lane	34%
Road Profile	Level	80%		No	64%		Stopped in Traffic Lane	-
	Other	20%		Unknown	6%		Passing Another Vehicle	-
Land Use	Rural	47%	**Violation**	Speeding	-		Parked in Travel Lane	-
	Urban	53%		Reckless	1%		Leaving a Parked Position	-
Day	Weekday	78%		None	46%		Entering a Parked Position	-
	Weekend	22%		Other	46%		Turning Right	-
Relation to Roadway	On Roadway	100%		Unknown	6%		Turning Left	-
	Shoulder/Parking Lane	0%	**Impairment**	Ill/Blackout	0.4%		Making U-turn	-
	Off Roadway	0%		Drowsy	2%		Backing Up	-
	Left Turn Lane	0%		None	91%		Negotiating a Curve	3%
	Unknown	0%		Other	1%		Changing Lanes	-
Relation to Junction	Non-Junction	16%		Unknown	5%		Merging	-
	Intersection	6%	**Gender**	Male	53%		Prior Corrective Action	-
	Intersection-Related	66%		Female	47%		Other	-
	Driveway/Alley	1%	**Age**	Younger <= 24	30%	**Driver Avoidance Maneuver**	Object in Road	-
	Entrance/Exit Ramp	6%		Middle = 25 to 64	65%		Poor Road Conditions	-
	Rail Grade Crossing	1%		Older >= 65	5%		Animal in Road	-
	Other/Unknown	4%					Vehicle in Road	19%
Posted Speed Limit (mph)	<= 20	1%					Non-Motorist in Road	-
	25	4%					Hit and Run	6%
	30	6%					No Driver Present	-
	35	22%					Other Avoidance Maneuver	-
	40	10%					Unknown	64%
	45	34%					None	12%
	50	4%					Phantom Vehicle	-
	>= 55	19%				**Corrective Action Attempted**	No Driver Present	-
Traffic Control Device	No Traffic Controls	21%					No Avoidance Maneuver	12%
	Traffic Signal	58%					Braking	15%
	Stop/Yield Sign	16%					Releasing Brakes	-
	Other	5%					Steering	4%
							Braked and Steered	-
							Accelerated	0.1%
							Accelerated and Steered	-
							Other	-
							Unknown	69%

Driver and vehicle statistics represent the striking light vehicle.

Following Vehicle Approaching Lead Vehicle Moving at Lower Constant Speed

Driving Environment

Lighting	Daylight	76%
	Dark Lighted	14%
	Dark	7%
	Dawn/Dusk	4%
Weather	Clear	85%
	Adverse	15%
Road Surface	Dry	79%
	Wet/Slippery	21%
Road Alignment	Straight	90%
	Curve	10%
Road Profile	Level	78%
	Other	22%
Land Use	Rural	43%
	Urban	57%
Day	Weekday	80%
	Weekend	20%
Relation to Roadway	On Roadway	99%
	Shoulder/Parking Lane	0.1%
	Off Roadway	1%
	Left Turn Lane	0.1%
	Unknown	0%
Relation to Junction	Non-Junction	61%
	Intersection	4%
	Intersection-Related	26%
	Driveway/Alley	2%
	Entrance/Exit Ramp	3%
	Rail Grade Crossing	0.3%
	Other/Unknown	3%
Posted Speed Limit (mph)	<= 20	0.4%
	25	6%
	30	7%
	35	20%
	40	10%
	45	21%
	50	4%
	>= 55	31%
Traffic Control Device	No Traffic Controls	69%
	Traffic Signal	20%
	Stop/Yield Sign	4%
	Other	7%

Driver

Alcohol	Yes	5%
	No	95%
Vision Obscured	No Obstruction	68%
	Vision Obscured	2%
	Unknown	29%
Driver Distracted	Inattention	25%
	Sleepy	2%
	Not Distracted	33%
	Unknown	41%
Speeding	Yes	36%
	No	59%
	Unknown	5%
Violation	Speeding	0.3%
	Reckless	1%
	None	45%
	Other	47%
	Unknown	7%
Impairment	Ill/Blackout	0.2%
	Drowsy	1%
	None	90%
	Other	3%
	Unknown	5%
Gender	Male	59%
	Female	41%
Age	Younger <= 24	36%
	Middle = 25 to 64	60%
	Older >= 65	4%

Vehicle

Contributing Factors	Yes	1%
	No	88%
	Unknown	11%
Rollover	Yes	1%
	No	99%
Pre-Event Movement	No Driver Present	-
	Going Straight	91%
	Decelerating in Traffic Lane	5%
	Accelerating in Traffic Lane	0.1%
	Starting in Traffic Lane	2%
	Stopped in Traffic Lane	0.1%
	Passing Another Vehicle	-
	Parked in Travel Lane	-
	Leaving a Parked Position	-
	Entering a Parked Position	-
	Turning Right	-
	Turning Left	0.1%
	Making U-turn	0.1%
	Backing Up	-
	Negotiating a Curve	1%
	Changing Lanes	-
	Merging	-
	Prior Corrective Action	-
	Other	-
Driver Avoidance Maneuver	Object in Road	0.002%
	Poor Road Conditions	-
	Animal in Road	-
	Vehicle in Road	27%
	Non-Motorist in Road	-
	Hit and Run	9%
	No Driver Present	-
	Other Avoidance Maneuver	-
	Unknown	56%
	None	9%
	Phantom Vehicle	0.01%
Corrective Action Attempted	No Driver Present	-
	No Avoidance Maneuver	7%
	Braking	23%
	Releasing Brakes	-
	Steering	3%
	Braked and Steered	2%
	Accelerated	0.2%
	Accelerated and Steered	-
	Other	0.3%
	Unknown	65%

Driver and vehicle statistics represent the striking light vehicle.

Following Vehicle Approaching a Decelerating Lead Vehicle

Driving Environment

Lighting	Daylight	84%
	Dark Lighted	9%
	Dark	4%
	Dawn/Dusk	3%
Weather	Clear	84%
	Adverse	16%
Road Surface	Dry	78%
	Wet/Slippery	22%
Road Alignment	Straight	92%
	Curve	8%
Road Profile	Level	78%
	Other	22%
Land Use	Rural	52%
	Urban	48%
Day	Weekday	83%
	Weekend	17%
Relation to Roadway	On Roadway	98%
	Shoulder/Parking Lane	0.1%
	Off Roadway	2%
	Left Turn Lane	0.02%
	Unknown	0%
Relation to Junction	Non-Junction	53%
	Intersection	4%
	Intersection-Related	30%
	Driveway/Alley	7%
	Entrance/Exit Ramp	4%
	Rail Grade Crossing	0.2%
	Other/Unknown	2%
Posted Speed Limit (mph)	<= 20	1%
	25	6%
	30	5%
	35	20%
	40	11%
	45	21%
	50	6%
	>= 55	30%
Traffic Control Device	No Traffic Controls	69%
	Traffic Signal	19%
	Stop/Yield Sign	4%
	Other	7%

Driver

Alcohol	Yes	2%
	No	98%
Vision Obscured	No Obstruction	77%
	Vision Obscured	1%
	Unknown	22%
Driver Distracted	Inattention	32%
	Sleepy	0.3%
	Not Distracted	29%
	Unknown	39%
Speeding	Yes	43%
	No	54%
	Unknown	3%
Violation	Speeding	0.03%
	Reckless	1%
	None	47%
	Other	49%
	Unknown	3%
Impairment	Ill/Blackout	0.01%
	Drowsy	0.3%
	None	95%
	Other	1%
	Unknown	3%
Gender	Male	60%
	Female	40%
Age	Younger <= 24	39%
	Middle = 25 to 64	56%
	Older >= 65	5%

Vehicle

Contributing Factors	Yes	1%
	No	93%
	Unknown	6%
Rollover	Yes	1%
	No	99%
Pre-Event Movement	No Driver Present	-
	Going Straight	84%
	Decelerating in Traffic Lane	11%
	Accelerating in Traffic Lane	0.3%
	Starting in Traffic Lane	4%
	Stopped in Traffic Lane	-
	Passing Another Vehicle	-
	Parked in Travel Lane	-
	Leaving a Parked Position	-
	Entering a Parked Position	-
	Turning Right	-
	Turning Left	-
	Making U-turn	-
	Backing Up	-
	Negotiating a Curve	1%
	Changing Lanes	-
	Merging	-
	Prior Corrective Action	-
	Other	-
Driver Avoidance Maneuver	Object in Road	0.1%
	Poor Road Conditions	0.001%
	Animal in Road	-
	Vehicle in Road	24%
	Non-Motorist in Road	-
	Hit and Run	4%
	No Driver Present	-
	Other Avoidance Maneuver	-
	Unknown	55%
	None	16%
	Phantom Vehicle	0.3%
Corrective Action Attempted	No Driver Present	-
	No Avoidance Maneuver	13%
	Braking	18%
	Releasing Brakes	-
	Steering	4%
	Braked and Steered	2%
	Accelerated	0.1%
	Accelerated and Steered	0.1%
	Other	0.1%
	Unknown	63%

Driver and vehicle statistics represent the striking light vehicle.

Following Vehicle Approaching a Stopped Lead Vehicle

Driving Environment

Lighting	Daylight	81%
	Dark Lighted	12%
	Dark	4%
	Dawn/Dusk	3%
Weather	Clear	85%
	Adverse	15%
Road Surface	Dry	79%
	Wet/Slippery	21%
Road Alignment	Straight	91%
	Curve	9%
Road Profile	Level	80%
	Other	20%
Land Use	Rural	49%
	Urban	51%
Day	Weekday	82%
	Weekend	18%
Relation to Roadway	On Roadway	99%
	Shoulder/Parking Lane	0.1%
	Off Roadway	0.5%
	Left Turn Lane	0.3%
	Unknown	0.03%
Relation to Junction	Non-Junction	34%
	Intersection	4%
	Intersection-Related	50%
	Driveway/Alley	3%
	Entrance/Exit Ramp	5%
	Rail Grade Crossing	0.4%
	Other/Unknown	4%
Posted Speed Limit (mph)	<= 20	1%
	25	8%
	30	7%
	35	27%
	40	13%
	45	24%
	50	5%
	>= 55	15%
Traffic Control Device	No Traffic Controls	45%
	Traffic Signal	39%
	Stop/Yield Sign	9%
	Other	7%

Driver

Alcohol	Yes	4%
	No	96%
Vision Obscured	No Obstruction	71%
	Vision Obscured	2%
	Unknown	27%
Driver Distracted	Inattention	37%
	Sleepy	1%
	Not Distracted	21%
	Unknown	41%
Speeding	Yes	35%
	No	61%
	Unknown	4%
Violation	Speeding	0.1%
	Reckless	1%
	None	43%
	Other	51%
	Unknown	5%
Impairment	Ill/Blackout	0.1%
	Drowsy	1%
	None	93%
	Other	3%
	Unknown	4%
Gender	Male	58%
	Female	42%
Age	Younger <= 24	35%
	Middle = 25 to 64	59%
	Older >= 65	6%

Vehicle

Contributing Factors	Yes	1%
	No	90%
	Unknown	9%
Rollover	Yes	0.1%
	No	100%
Pre-Event Movement	No Driver Present	-
	Going Straight	77%
	Decelerating in Traffic Lane	12%
	Accelerating in Traffic Lane	1%
	Starting in Traffic Lane	8%
	Stopped in Traffic Lane	0.04%
	Passing Another Vehicle	-
	Parked in Travel Lane	-
	Leaving a Parked Position	-
	Entering a Parked Position	-
	Turning Right	0.05%
	Turning Left	-
	Making U-turn	-
	Backing Up	-
	Negotiating a Curve	2%
	Changing Lanes	-
	Merging	-
	Prior Corrective Action	-
	Other	-
Driver Avoidance Maneuver	Object in Road	0.01%
	Poor Road Conditions	0.04%
	Animal in Road	-
	Vehicle in Road	18%
	Non-Motorist in Road	-
	Hit and Run	6%
	No Driver Present	-
	Other Avoidance Maneuver	-
	Unknown	59%
	None	16%
	Phantom Vehicle	0.01%
Corrective Action Attempted	No Driver Present	-
	No Avoidance Maneuver	13%
	Braking	16%
	Releasing Brakes	-
	Steering	3%
	Braked and Steered	1%
	Accelerated	0.4%
	Accelerated and Steered	0.02%
	Other	0.1%
	Unknown	66%

Driver and vehicle statistics represent the striking light vehicle.

Left Turn Across Path From Opposite Directions at Signalized Junctions

Driving Environment

Lighting	Daylight	24%
	Dark Lighted	8%
	Dark	58%
	Dawn/Dusk	9%
Weather	Clear	91%
	Adverse	9%
Road Surface	Dry	82%
	Wet/Slippery	18%
Road Alignment	Straight	89%
	Curve	11%
Road Profile	Level	74%
	Other	26%
Land Use	Rural	79%
	Urban	21%
Day	Weekday	70%
	Weekend	30%
Relation to Roadway	On Roadway	90%
	Shoulder/Parking Lane	0.4%
	Off Roadway	9%
	Left Turn Lane	-
	Unknown	0.1%
Relation to Junction	Non-Junction	97%
	Intersection	1%
	Intersection-Related	1%
	Driveway/Alley	-
	Entrance/Exit Ramp	1%
	Rail Grade Crossing	-
	Other/Unknown	1%
Posted Speed Limit (mph)	<= 20	1%
	25	5%
	30	2%
	35	8%
	40	4%
	45	12%
	50	5%
	>= 55	62%
Traffic Control Device	No Traffic Controls	91%
	Traffic Signal	1%
	Stop/Yield Sign	0.02%
	Other	8%

Driver

Alcohol	Yes	1%
	No	99%
Vision Obscured	No Obstruction	87%
	Vision Obscured	1%
	Unknown	13%
Driver Distracted	Inattention	1%
	Sleepy	-
	Not Distracted	74%
	Unknown	25%
Speeding	Yes	2%
	No	97%
	Unknown	1%
Violation	Speeding	-
	Reckless	0.1%
	None	97%
	Other	3%
	Unknown	0.1%
Impairment	Ill/Blackout	-
	Drowsy	-
	None	98%
	Other	0.3%
	Unknown	2%
Gender	Male	61%
	Female	39%
Age	Younger <= 24	20%
	Middle = 25 to 64	74%
	Older >= 65	5%

Vehicle

Contributing Factors	Yes	0.1%
	No	96%
	Unknown	4%
Rollover	Yes	2%
	No	98%
Pre-Event Movement	No Driver Present	-
	Going Straight	94%
	Decelerating in Traffic Lane	0.4%
	Accelerating in Traffic Lane	0.1%
	Starting in Traffic Lane	0.1%
	Stopped in Traffic Lane	0.3%
	Passing Another Vehicle	-
	Parked in Travel Lane	-
	Leaving a Parked Position	-
	Entering a Parked Position	-
	Turning Right	-
	Turning Left	-
	Making U-turn	-
	Backing Up	-
	Negotiating a Curve	5%
	Changing Lanes	-
	Merging	-
	Prior Corrective Action	-
	Other	-
Driver Avoidance Maneuver	Object in Road	-
	Poor Road Conditions	-
	Animal in Road	17%
	Vehicle in Road	0.03%
	Non-Motorist in Road	-
	Hit and Run	0.3%
	No Driver Present	-
	Other Avoidance Maneuver	-
	Unknown	69%
	None	13%
	Phantom Vehicle	0.1%
Corrective Action Attempted	No Driver Present	-
	No Avoidance Maneuver	8%
	Braking	4%
	Releasing Brakes	-
	Steering	10%
	Braked and Steered	1%
	Accelerated	-
	Accelerated and Steered	0.01%
	Other	1%
	Unknown	76%

Driver and vehicle statistics represent the light vehicle turning left.

Vehicle Turning Right at Signalized Junctions

Driving Environment

Lighting	Daylight	71%
	Dark Lighted	24%
	Dark	3%
	Dawn/Dusk	2%
Weather	Clear	80%
	Adverse	20%
Road Surface	Dry	73%
	Wet/Slippery	27%
Road Alignment	Straight	93%
	Curve	7%
Road Profile	Level	81%
	Other	19%
Land Use	Rural	46%
	Urban	54%
Day	Weekday	78%
	Weekend	22%
Relation to Roadway	On Roadway	99%
	Shoulder/Parking Lane	-
	Off Roadway	-
	Left Turn Lane	1%
	Unknown	-
Relation to Junction	Non-Junction	-
	Intersection	54%
	Intersection-Related	37%
	Driveway/Alley	4%
	Entrance/Exit Ramp	2%
	Rail Grade Crossing	-
	Other/Unknown	3%
Posted Speed Limit (mph)	<= 20	1%
	25	10%
	30	10%
	35	29%
	40	16%
	45	26%
	50	4%
	>= 55	6%
Traffic Control Device	No Traffic Controls	-
	Traffic Signal	100%
	Stop/Yield Sign	-
	Other	-

Driver

Alcohol	Yes	5%
	No	95%
Vision Obscured	No Obstruction	60%
	Vision Obscured	5%
	Unknown	34%
Driver Distracted	Inattention	16%
	Sleepy	0.02%
	Not Distracted	28%
	Unknown	56%
Speeding	Yes	8%
	No	83%
	Unknown	8%
Violation	Speeding	-
	Reckless	0.03%
	None	53%
	Other	38%
	Unknown	10%
Impairment	Ill/Blackout	0.02%
	Drowsy	-
	None	89%
	Other	2%
	Unknown	9%
Gender	Male	58%
	Female	42%
Age	Younger <= 24	35%
	Middle = 25 to 64	48%
	Older >= 65	16%

Vehicle

Contributing Factors	Yes	1%
	No	82%
	Unknown	18%
Rollover	Yes	-
	No	100%
Pre-Event Movement	No Driver Present	-
	Going Straight	1%
	Decelerating in Traffic Lane	-
	Accelerating in Traffic Lane	-
	Starting in Traffic Lane	2%
	Stopped in Traffic Lane	-
	Passing Another Vehicle	-
	Parked in Travel Lane	-
	Leaving a Parked Position	-
	Entering a Parked Position	-
	Turning Right	97%
	Turning Left	-
	Making U-turn	-
	Backing Up	-
	Negotiating a Curve	-
	Changing Lanes	-
	Merging	-
	Prior Corrective Action	-
	Other	-
Driver Avoidance Maneuver	Object in Road	-
	Poor Road Conditions	-
	Animal in Road	-
	Vehicle in Road	1%
	Non-Motorist in Road	-
	Hit and Run	12%
	No Driver Present	-
	Other Avoidance Maneuver	-
	Unknown	69%
	None	19%
	Phantom Vehicle	0.01%
Corrective Action Attempted	No Driver Present	-
	No Avoidance Maneuver	15%
	Braking	1%
	Releasing Brakes	-
	Steering	0.03%
	Braked and Steered	1%
	Accelerated	-
	Accelerated and Steered	-
	Other	-
	Unknown	84%

Driver and vehicle statistics represent the light vehicle turning right.

Left Turn Across Path From Opposite Directions at Non-Signalized Junctions

Driving Environment

Lighting	Daylight	80%
	Dark Lighted	12%
	Dark	4%
	Dawn/Dusk	3%
Weather	Clear	89%
	Adverse	11%
Road Surface	Dry	84%
	Wet/Slippery	16%
Road Alignment	Straight	93%
	Curve	7%
Road Profile	Level	80%
	Other	20%
Land Use	Rural	50%
	Urban	50%
Day	Weekday	83%
	Weekend	17%
Relation to Roadway	On Roadway	99%
	Shoulder/Parking Lane	1%
	Off Roadway	0.1%
	Left Turn Lane	0.1%
	Unknown	-
Relation to Junction	Non-Junction	0.3%
	Intersection	56%
	Intersection-Related	2%
	Driveway/Alley	40%
	Entrance/Exit Ramp	0.4%
	Rail Grade Crossing	-
	Other/Unknown	2%
Posted Speed Limit (mph)	<= 20	2%
	25	11%
	30	11%
	35	32%
	40	15%
	45	17%
	50	3%
	>= 55	9%
Traffic Control Device	No Traffic Controls	80%
	Traffic Signal	2%
	Stop/Yield Sign	10%
	Other	9%

Driver

Alcohol	Yes	3%
	No	97%
Vision Obscured	No Obstruction	58%
	Vision Obscured	16%
	Unknown	26%
Driver Distracted	Inattention	26%
	Sleepy	0.01%
	Not Distracted	33%
	Unknown	41%
Speeding	Yes	1%
	No	97%
	Unknown	2%
Violation	Speeding	0.04%
	Reckless	0.2%
	None	46%
	Other	51%
	Unknown	3%
Impairment	Ill/Blackout	0.1%
	Drowsy	0.01%
	None	96%
	Other	2%
	Unknown	2%
Gender	Male	56%
	Female	44%
Age	Younger <= 24	30%
	Middle = 25 to 64	55%
	Older >= 65	15%

Vehicle

Contributing Factors	Yes	0.2%
	No	95%
	Unknown	5%
Rollover	Yes	1%
	No	99%
Pre-Event Movement	No Driver Present	-
	Going Straight	1%
	Decelerating in Traffic Lane	-
	Accelerating in Traffic Lane	0.03%
	Starting in Traffic Lane	0.2%
	Stopped in Traffic Lane	0.1%
	Passing Another Vehicle	-
	Parked in Travel Lane	-
	Leaving a Parked Position	0.1%
	Entering a Parked Position	-
	Turning Right	0.1%
	Turning Left	98%
	Making U-turn	0.1%
	Backing Up	-
	Negotiating a Curve	0.2%
	Changing Lanes	-
	Merging	0.1%
	Prior Corrective Action	-
	Other	0.1%
Driver Avoidance Maneuver	Object in Road	-
	Poor Road Conditions	-
	Animal in Road	-
	Vehicle in Road	3%
	Non-Motorist in Road	-
	Hit and Run	4%
	No Driver Present	-
	Other Avoidance Maneuver	-
	Unknown	73%
	None	21%
	Phantom Vehicle	-
Corrective Action Attempted	No Driver Present	-
	No Avoidance Maneuver	19%
	Braking	1%
	Releasing Brakes	-
	Steering	1%
	Braked and Steered	0.01%
	Accelerated	1%
	Accelerated and Steered	0.02%
	Other	0.1%
	Unknown	79%

Driver and vehicle statistics represent the light vehicle turning left.

Straight Crossing Paths at Non-Signalized Junctions

Driving Environment

Category	Value	%
Lighting	Daylight	81%
	Dark Lighted	11%
	Dark	4%
	Dawn/Dusk	3%
Weather	Clear	86%
	Adverse	14%
Road Surface	Dry	79%
	Wet/Slippery	21%
Road Alignment	Straight	94%
	Curve	6%
Road Profile	Level	84%
	Other	16%
Land Use	Rural	49%
	Urban	51%
Day	Weekday	77%
	Weekend	23%
Relation to Roadway	On Roadway	99%
	Shoulder/Parking Lane	0.2%
	Off Roadway	1%
	Left Turn Lane	0.04%
	Unknown	-
Relation to Junction	Non-Junction	0.4%
	Intersection	87%
	Intersection-Related	1%
	Driveway/Alley	10%
	Entrance/Exit Ramp	0.2%
	Rail Grade Crossing	0.1%
	Other/Unknown	0.2%
Posted Speed Limit (mph)	<= 20	3%
	25	27%
	30	16%
	35	25%
	40	7%
	45	11%
	50	3%
	>= 55	9%
Traffic Control Device	No Traffic Controls	17%
	Traffic Signal	3%
	Stop/Yield Sign	77%
	Other	3%

Driver

Category	Value	%
Alcohol	Yes	2%
	No	98%
Vision Obscured	No Obstruction	68%
	Vision Obscured	6%
	Unknown	26%
Driver Distracted	Inattention	14%
	Sleepy	0.1%
	Not Distracted	45%
	Unknown	40%
Speeding	Yes	2%
	No	96%
	Unknown	2%
Violation	Speeding	0.1%
	Reckless	0.3%
	None	66%
	Other	32%
	Unknown	2%
Impairment	Ill/Blackout	0.03%
	Drowsy	0.05%
	None	96%
	Other	1%
	Unknown	3%
Gender	Male	52%
	Female	48%
Age	Younger <= 24	30%
	Middle = 25 to 64	60%
	Older >= 65	11%

Vehicle

Category	Value	%
Contributing Factors	Yes	0.5%
	No	93%
	Unknown	6%
Rollover	Yes	1%
	No	99%
Pre-Event Movement	No Driver Present	0.1%
	Going Straight	76%
	Decelerating in Traffic Lane	1%
	Accelerating in Traffic Lane	0.2%
	Starting in Traffic Lane	20%
	Stopped in Traffic Lane	0.5%
	Passing Another Vehicle	0.2%
	Parked in Travel Lane	0.04%
	Leaving a Parked Position	1%
	Entering a Parked Position	0.01%
	Turning Right	-
	Turning Left	0.1%
	Making U-turn	0.1%
	Backing Up	-
	Negotiating a Curve	0.3%
	Changing Lanes	0.1%
	Merging	0.05%
	Prior Corrective Action	0.2%
	Other	1%
Driver Avoidance Maneuver	Object in Road	-
	Poor Road Conditions	-
	Animal in Road	-
	Vehicle in Road	9%
	Non-Motorist in Road	-
	Hit and Run	3%
	No Driver Present	0.1%
	Other Avoidance Maneuver	-
	Unknown	70%
	None	19%
	Phantom Vehicle	0.05%
Corrective Action Attempted	No Driver Present	0.1%
	No Avoidance Maneuver	15%
	Braking	5%
	Releasing Brakes	-
	Steering	2%
	Braked and Steered	1%
	Accelerated	0.4%
	Accelerated and Steered	-
	Other	0.2%
	Unknown	76%

Driver and vehicle statistics represent all light vehicles involved.

Vehicle(s) Turning at Non-Signalized Junctions

Driving Environment

Lighting	Daylight	79%
	Dark Lighted	12%
	Dark	5%
	Dawn/Dusk	4%
Weather	Clear	87%
	Adverse	13%
Road Surface	Dry	81%
	Wet/Slippery	19%
Road Alignment	Straight	93%
	Curve	7%
Road Profile	Level	80%
	Other	20%
Land Use	Rural	51%
	Urban	49%
Day	Weekday	80%
	Weekend	20%
Relation to Roadway	On Roadway	97%
	Shoulder/Parking Lane	0.2%
	Off Roadway	2%
	Left Turn Lane	1%
	Unknown	0.04%
Relation to Junction	Non-Junction	1%
	Intersection	47%
	Intersection-Related	9%
	Driveway/Alley	40%
	Entrance/Exit Ramp	1%
	Rail Grade Crossing	0.03%
	Other/Unknown	1%
Posted Speed Limit (mph)	<= 20	3%
	25	15%
	30	9%
	35	28%
	40	13%
	45	19%
	50	4%
	>= 55	9%
Traffic Control Device	No Traffic Controls	47%
	Traffic Signal	1%
	Stop/Yield Sign	46%
	Other	6%

Driver

Alcohol	Yes	2%
	No	98%
Vision Obscured	No Obstruction	62%
	Vision Obscured	12%
	Unknown	26%
Driver Distracted	Inattention	26%
	Sleepy	-
	Not Distracted	31%
	Unknown	43%
Speeding	Yes	2%
	No	95%
	Unknown	3%
Violation	Speeding	0.2%
	Reckless	0.3%
	None	48%
	Other	48%
	Unknown	4%
Impairment	Ill/Blackout	0.1%
	Drowsy	0.03%
	None	95%
	Other	1%
	Unknown	3%
Gender	Male	53%
	Female	47%
Age	Younger <= 24	35%
	Middle = 25 to 64	52%
	Older >= 65	12%

Vehicle

Contributing Factors	Yes	0.1%
	No	92%
	Unknown	8%
Rollover	Yes	0.3%
	No	100%
Pre-Event Movement	No Driver Present	0.02%
	Going Straight	1%
	Decelerating in Traffic Lane	0.02%
	Accelerating in Traffic Lane	0.03%
	Starting in Traffic Lane	1%
	Stopped in Traffic Lane	0.1%
	Passing Another Vehicle	-
	Parked in Travel Lane	-
	Leaving a Parked Position	-
	Entering a Parked Position	-
	Turning Right	22%
	Turning Left	74%
	Making U-turn	0.5%
	Backing Up	-
	Negotiating a Curve	0.1%
	Changing Lanes	0.03%
	Merging	-
	Prior Corrective Action	0.3%
	Other	0.1%
Driver Avoidance Maneuver	Object in Road	-
	Poor Road Conditions	-
	Animal in Road	-
	Vehicle in Road	3%
	Non-Motorist in Road	-
	Hit and Run	5%
	No Driver Present	0.02%
	Other Avoidance Maneuver	-
	Unknown	68%
	None	23%
	Phantom Vehicle	0.1%
Corrective Action Attempted	No Driver Present	0.02%
	No Avoidance Maneuver	20%
	Braking	1%
	Releasing Brakes	0.1%
	Steering	2%
	Braked and Steered	0.3%
	Accelerated	0.2%
	Accelerated and Steered	-
	Other	0.1%
	Unknown	76%

Driver and vehicle statistics represent all turning light vehicles involved.

Vehicle Taking Evasive Action With Prior Vehicle Maneuver

Driving Environment

Lighting	Daylight	66%
	Dark Lighted	24%
	Dark	6%
	Dawn/Dusk	3%
Weather	Clear	86%
	Adverse	14%
Road Surface	Dry	77%
	Wet/Slippery	23%
Road Alignment	Straight	86%
	Curve	14%
Road Profile	Level	85%
	Other	15%
Land Use	Rural	30%
	Urban	70%
Day	Weekday	68%
	Weekend	32%
Relation to Roadway	On Roadway	66%
	Shoulder/Parking Lane	4%
	Off Roadway	28%
	Left Turn Lane	-
	Unknown	2%
Relation to Junction	Non-Junction	34%
	Intersection	7%
	Intersection-Related	34%
	Driveway/Alley	11%
	Entrance/Exit Ramp	8%
	Rail Grade Crossing	-
	Other/Unknown	7%
Posted Speed Limit (mph)	<= 20	6%
	25	10%
	30	7%
	35	28%
	40	7%
	45	21%
	50	6%
	>= 55	14%
Traffic Control Device	No Traffic Controls	58%
	Traffic Signal	25%
	Stop/Yield Sign	10%
	Other	7%

Driver

Alcohol	Yes	4%
	No	96%
Vision Obscured	No Obstruction	61%
	Vision Obscured	2%
	Unknown	37%
Driver Distracted	Inattention	12%
	Sleepy	-
	Not Distracted	41%
	Unknown	47%
Speeding	Yes	7%
	No	91%
	Unknown	1%
Violation	Speeding	-
	Reckless	1%
	None	75%
	Other	19%
	Unknown	5%
Impairment	Ill/Blackout	-
	Drowsy	-
	None	95%
	Other	0.3%
	Unknown	5%
Gender	Male	68%
	Female	32%
Age	Younger <= 24	32%
	Middle = 25 to 64	60%
	Older >= 65	8%

Vehicle

Contributing Factors	Yes	0.1%
	No	95%
	Unknown	5%
Rollover	Yes	1%
	No	99%
Pre-Event Movement	No Driver Present	1%
	Going Straight	19%
	Decelerating in Traffic Lane	3%
	Accelerating in Traffic Lane	-
	Starting in Traffic Lane	1%
	Stopped in Traffic Lane	1%
	Passing Another Vehicle	4%
	Parked in Travel Lane	-
	Leaving a Parked Position	4%
	Entering a Parked Position	-
	Turning Right	5%
	Turning Left	10%
	Making U-turn	1%
	Backing Up	7%
	Negotiating a Curve	-
	Changing Lanes	4%
	Merging	2%
	Prior Corrective Action	9%
	Other	29%
Driver Avoidance Maneuver	Object in Road	-
	Poor Road Conditions	-
	Animal in Road	-
	Vehicle in Road	25%
	Non-Motorist in Road	-
	Hit and Run	5%
	No Driver Present	1%
	Other Avoidance Maneuver	-
	Unknown	52%
	None	14%
	Phantom Vehicle	3%
Corrective Action Attempted	No Driver Present	1%
	No Avoidance Maneuver	12%
	Braking	5%
	Releasing Brakes	-
	Steering	21%
	Braked and Steered	1%
	Accelerated	1%
	Accelerated and Steered	-
	Other	0.1%
	Unknown	59%

Driver and vehicle statistics represent all light vehicles involved.

Vehicle Taking Evasive Action Without Prior Vehicle Maneuver

Driving Environment

Category	Value	%
Lighting	Daylight	72%
	Dark Lighted	17%
	Dark	8%
	Dawn/Dusk	3%
Weather	Clear	86%
	Adverse	14%
Road Surface	Dry	78%
	Wet/Slippery	22%
Road Alignment	Straight	86%
	Curve	14%
Road Profile	Level	77%
	Other	23%
Land Use	Rural	44%
	Urban	56%
Day	Weekday	78%
	Weekend	22%
Relation to Roadway	On Roadway	65%
	Shoulder/Parking Lane	3%
	Off Roadway	31%
	Left Turn Lane	0.005%
	Unknown	0%
Relation to Junction	Non-Junction	45%
	Intersection	14%
	Intersection-Related	13%
	Driveway/Alley	17%
	Entrance/Exit Ramp	1%
	Rail Grade Crossing	-
	Other/Unknown	9%
Posted Speed Limit (mph)	<= 20	3%
	25	12%
	30	9%
	35	25%
	40	11%
	45	17%
	50	3%
	>= 55	20%
Traffic Control Device	No Traffic Controls	70%
	Traffic Signal	15%
	Stop/Yield Sign	9%
	Other	6%

Driver

Category	Value	%
Alcohol	Yes	3%
	No	97%
Vision Obscured	No Obstruction	70%
	Vision Obscured	4%
	Unknown	26%
Driver Distracted	Inattention	14%
	Sleepy	-
	Not Distracted	44%
	Unknown	42%
Speeding	Yes	6%
	No	90%
	Unknown	4%
Violation	Speeding	-
	Reckless	1%
	None	69%
	Other	25%
	Unknown	5%
Impairment	Ill/Blackout	0.02%
	Drowsy	-
	None	94%
	Other	2%
	Unknown	5%
Gender	Male	58%
	Female	42%
Age	Younger <= 24	33%
	Middle = 25 to 64	59%
	Older >= 65	8%

Vehicle

Category	Value	%
Contributing Factors	Yes	1%
	No	92%
	Unknown	7%
Rollover	Yes	3%
	No	97%
Pre-Event Movement	No Driver Present	1%
	Going Straight	55%
	Decelerating in Traffic Lane	3%
	Accelerating in Traffic Lane	0.01%
	Starting in Traffic Lane	1%
	Stopped in Traffic Lane	3%
	Passing Another Vehicle	1%
	Parked in Travel Lane	-
	Leaving a Parked Position	4%
	Entering a Parked Position	0.01%
	Turning Right	-
	Turning Left	1%
	Making U-turn	11%
	Backing Up	-
	Negotiating a Curve	3%
	Changing Lanes	0.1%
	Merging	0.02%
	Prior Corrective Action	-
	Other	17%
Driver Avoidance Maneuver	Object in Road	2%
	Poor Road Conditions	-
	Animal in Road	-
	Vehicle in Road	21%
	Non-Motorist in Road	-
	Hit and Run	5%
	No Driver Present	1%
	Other Avoidance Maneuver	0.2%
	Unknown	58%
	None	9%
	Phantom Vehicle	4%
Corrective Action Attempted	No Driver Present	1%
	No Avoidance Maneuver	7%
	Braking	5%
	Releasing Brakes	-
	Steering	18%
	Braked and Steered	2%
	Accelerated	-
	Accelerated and Steered	0.3%
	Other	1%
	Unknown	66%

Driver and vehicle statistics represent all light vehicles involved.

Non-Collision Incident

Driving Environment

Lighting	Daylight	80%
	Dark Lighted	6%
	Dark	12%
	Dawn/Dusk	2%
Weather	Clear	94%
	Adverse	6%
Road Surface	Dry	89%
	Wet/Slippery	11%
Road Alignment	Straight	89%
	Curve	11%
Road Profile	Level	74%
	Other	26%
Land Use	Rural	63%
	Urban	37%
Day	Weekday	75%
	Weekend	25%
Relation to Roadway	On Roadway	90%
	Shoulder/Parking Lane	4%
	Off Roadway	6%
	Left Turn Lane	-
	Unknown	0.4%
Relation to Junction	Non-Junction	89%
	Intersection	1%
	Intersection-Related	6%
	Driveway/Alley	1%
	Entrance/Exit Ramp	2%
	Rail Grade Crossing	-
	Other/Unknown	0.1%
Posted Speed Limit (mph)	<= 20	1%
	25	6%
	30	4%
	35	10%
	40	2%
	45	10%
	50	4%
	>= 55	63%
Traffic Control Device	No Traffic Controls	84%
	Traffic Signal	5%
	Stop/Yield Sign	2%
	Other	9%

Driver

Alcohol	Yes	3%
	No	97%
Vision Obscured	No Obstruction	77%
	Vision Obscured	0.2%
	Unknown	23%
Driver Distracted	Inattention	4%
	Sleepy	-
	Not Distracted	55%
	Unknown	41%
Speeding	Yes	4%
	No	87%
	Unknown	9%
Violation	Speeding	0.1%
	Reckless	0.04%
	None	69%
	Other	25%
	Unknown	6%
Impairment	Ill/Blackout	-
	Drowsy	-
	None	95%
	Other	2%
	Unknown	4%
Gender	Male	70%
	Female	30%
Age	Younger <= 24	24%
	Middle = 25 to 64	72%
	Older >= 65	4%

Vehicle

Contributing Factors	Yes	42%
	No	48%
	Unknown	10%
Rollover	Yes	1%
	No	99%
Pre-Event Movement	No Driver Present	-
	Going Straight	80%
	Decelerating in Traffic Lane	1%
	Accelerating in Traffic Lane	-
	Starting in Traffic Lane	0.1%
	Stopped in Traffic Lane	1%
	Passing Another Vehicle	0.1%
	Parked in Travel Lane	-
	Leaving a Parked Position	2%
	Entering a Parked Position	-
	Turning Right	2%
	Turning Left	2%
	Making U-turn	0.1%
	Backing Up	1%
	Negotiating a Curve	6%
	Changing Lanes	1%
	Merging	0.1%
	Prior Corrective Action	-
	Other	2%
Driver Avoidance Maneuver	Object in Road	2%
	Poor Road Conditions	-
	Animal in Road	-
	Vehicle in Road	0.04%
	Non-Motorist in Road	-
	Hit and Run	5%
	No Driver Present	-
	Other Avoidance Maneuver	0.01%
	Unknown	24%
	None	69%
	Phantom Vehicle	-
Corrective Action Attempted	No Driver Present	-
	No Avoidance Maneuver	64%
	Braking	2%
	Releasing Brakes	-
	Steering	2%
	Braked and Steered	-
	Accelerated	0.1%
	Accelerated and Steered	-
	Other	1%
	Unknown	32%

Driver and vehicle statistics represent all light vehicles involved.

Vehicle Contacting Object With Prior Vehicle Maneuver

Driving Environment *Driver* *Vehicle*

Category	Value	%		Category	Value	%		Category	Value	%
Lighting	Daylight	46%		Alcohol	Yes	21%		Contributing Factors	Yes	0.5%
	Dark Lighted	34%			No	79%			No	25%
	Dark	16%		Vision Obscured	No Obstruction	19%			Unknown	74%
	Dawn/Dusk	4%			Vision Obscured	1%		Rollover	Yes	1%
Weather	Clear	87%			Unknown	80%			No	99%
	Adverse	13%		Driver Distracted	Inattention	4%		Pre-Event Movement	No Driver Present	-
Road Surface	Dry	64%			Sleepy	-			Going Straight	-
	Wet/Slippery	36%			Not Distracted	6%			Decelerating in Traffic Lane	-
Road Alignment	Straight	89%			Unknown	90%			Accelerating in Traffic Lane	-
	Curve	11%		Speeding	Yes	7%			Starting in Traffic Lane	-
Road Profile	Level	82%			No	26%			Stopped in Traffic Lane	-
	Other	18%			Unknown	67%			Passing Another Vehicle	8%
Land Use	Rural	34%		Violation	Speeding	-			Parked in Travel Lane	-
	Urban	66%			Reckless	1%			Leaving a Parked Position	20%
Day	Weekday	65%			None	21%			Entering a Parked Position	6%
	Weekend	35%			Other	14%			Turning Right	9%
Relation to Roadway	On Roadway	3%			Unknown	65%			Turning Left	3%
	Shoulder/Parking Lane	64%		Impairment	Ill/Blackout	-			Making U-turn	5%
	Off Roadway	30%			Drowsy	-			Backing Up	2%
	Left Turn Lane	-			None	63%			Negotiating a Curve	-
	Unknown	4%			Other	6%			Changing Lanes	1%
Relation to Junction	Non-Junction	70%			Unknown	32%			Merging	0.4%
	Intersection	1%		Gender	Male	62%			Prior Corrective Action	1%
	Intersection-Related	13%			Female	38%			Other	46%
	Driveway/Alley	9%		Age	Younger <= 24	71%		Driver Avoidance Maneuver	Object in Road	1%
	Entrance/Exit Ramp	1%			Middle = 25 to 64	29%			Poor Road Conditions	-
	Rail Grade Crossing	0.02%			Older >= 65	0.1%			Animal in Road	-
	Other/Unknown	6%							Vehicle in Road	1%
Posted Speed Limit (mph)	<= 20	8%							Non-Motorist in Road	-
	25	39%							Hit and Run	75%
	30	13%							No Driver Present	-
	35	16%							Other Avoidance Maneuver	1%
	40	3%							Unknown	18%
	45	5%							None	3%
	50	4%							Phantom Vehicle	0.5%
	>= 55	11%						Corrective Action Attempted	No Driver Present	-
Traffic Control Device	No Traffic Controls	82%							No Avoidance Maneuver	2%
	Traffic Signal	4%							Braking	0.3%
	Stop/Yield Sign	1%							Releasing Brakes	-
	Other	14%							Steering	2%
									Braked and Steered	-
									Accelerated	0.1%
									Accelerated and Steered	-
									Other	1%
									Unknown	95%

Vehicle Contacting Object Without Prior Vehicle Maneuver

Driving Environment

Lighting	Daylight	49%
	Dark Lighted	17%
	Dark	29%
	Dawn/Dusk	5%
Weather	Clear	86%
	Adverse	14%
Road Surface	Dry	77%
	Wet/Slippery	23%
Road Alignment	Straight	82%
	Curve	18%
Road Profile	Level	74%
	Other	26%
Land Use	Rural	53%
	Urban	47%
Day	Weekday	69%
	Weekend	31%
Relation to Roadway	On Roadway	54%
	Shoulder/Parking Lane	14%
	Off Roadway	30%
	Left Turn Lane	-
	Unknown	2%
Relation to Junction	Non-Junction	86%
	Intersection	2%
	Intersection-Related	4%
	Driveway/Alley	0.3%
	Entrance/Exit Ramp	3%
	Rail Grade Crossing	4%
	Other/Unknown	2%
Posted Speed Limit (mph)	<= 20	2%
	25	16%
	30	9%
	35	13%
	40	4%
	45	12%
	50	3%
	>= 55	41%
Traffic Control Device	No Traffic Controls	82%
	Traffic Signal	2%
	Stop/Yield Sign	1%
	Other	15%

Driver

Alcohol	Yes	11%
	No	89%
Vision Obscured	No Obstruction	62%
	Vision Obscured	4%
	Unknown	34%
Driver Distracted	Inattention	13%
	Sleepy	2%
	Not Distracted	40%
	Unknown	45%
Speeding	Yes	10%
	No	72%
	Unknown	18%
Violation	Speeding	0.4%
	Reckless	0.5%
	None	68%
	Other	14%
	Unknown	17%
Impairment	Ill/Blackout	-
	Drowsy	2%
	None	81%
	Other	2%
	Unknown	14%
Gender	Male	59%
	Female	41%
Age	Younger <= 24	36%
	Middle = 25 to 64	57%
	Older >= 65	7%

Vehicle

Contributing Factors	Yes	1%
	No	80%
	Unknown	19%
Rollover	Yes	5%
	No	95%
Pre-Event Movement	No Driver Present	-
	Going Straight	90%
	Decelerating in Traffic Lane	0.4%
	Accelerating in Traffic Lane	0.01%
	Starting in Traffic Lane	1%
	Stopped in Traffic Lane	1%
	Passing Another Vehicle	-
	Parked in Travel Lane	-
	Leaving a Parked Position	-
	Entering a Parked Position	-
	Turning Right	-
	Turning Left	-
	Making U-turn	-
	Backing Up	-
	Negotiating a Curve	8%
	Changing Lanes	-
	Merging	-
	Prior Corrective Action	-
	Other	-
Driver Avoidance Maneuver	Object in Road	8%
	Poor Road Conditions	0.4%
	Animal in Road	0.2%
	Vehicle in Road	1%
	Non-Motorist in Road	-
	Hit and Run	20%
	No Driver Present	-
	Other Avoidance Maneuver	0.2%
	Unknown	46%
	None	24%
	Phantom Vehicle	0.4%
Corrective Action Attempted	No Driver Present	-
	No Avoidance Maneuver	16%
	Braking	2%
	Releasing Brakes	-
	Steering	7%
	Braked and Steered	0.2%
	Accelerated	-
	Accelerated and Steered	-
	Other	1%
	Unknown	75%

Other

Driving Environment

Category	Value	%
Lighting	Daylight	67%
	Dark Lighted	17%
	Dark	13%
	Dawn/Dusk	3%
Weather	Clear	89%
	Adverse	11%
Road Surface	Dry	75%
	Wet/Slippery	25%
Road Alignment	Straight	86%
	Curve	14%
Road Profile	Level	72%
	Other	28%
Land Use	Rural	43%
	Urban	57%
Day	Weekday	69%
	Weekend	31%
Relation to Roadway	On Roadway	81%
	Shoulder/Parking Lane	7%
	Off Roadway	9%
	Left Turn Lane	1%
	Unknown	3%
Relation to Junction	Non-Junction	41%
	Intersection	16%
	Intersection-Related	15%
	Driveway/Alley	18%
	Entrance/Exit Ramp	1%
	Rail Grade Crossing	0.4%
	Other/Unknown	8%
Posted Speed Limit (mph)	<= 20	9%
	25	24%
	30	9%
	35	20%
	40	6%
	45	12%
	50	4%
	>= 55	16%
Traffic Control Device	No Traffic Controls	64%
	Traffic Signal	17%
	Stop/Yield Sign	5%
	Other	14%

Driver

Category	Value	%
Alcohol	Yes	4%
	No	96%
Vision Obscured	No Obstruction	53%
	Vision Obscured	1%
	Unknown	46%
Driver Distracted	Inattention	11%
	Sleepy	1%
	Not Distracted	36%
	Unknown	53%
Speeding	Yes	3%
	No	89%
	Unknown	8%
Violation	Speeding	-
	Reckless	0.2%
	None	64%
	Other	19%
	Unknown	17%
Impairment	Ill/Blackout	-
	Drowsy	1%
	None	88%
	Other	1%
	Unknown	9%
Gender	Male	59%
	Female	41%
Age	Younger <= 24	30%
	Middle = 25 to 64	62%
	Older >= 65	8%

Vehicle

Category	Value	%
Contributing Factors	Yes	2%
	No	84%
	Unknown	14%
Rollover	Yes	6%
	No	94%
Pre-Event Movement	No Driver Present	11%
	Going Straight	28%
	Decelerating in Traffic Lane	0.01%
	Accelerating in Traffic Lane	-
	Starting in Traffic Lane	0.4%
	Stopped in Traffic Lane	4%
	Passing Another Vehicle	2%
	Parked in Travel Lane	0.01%
	Leaving a Parked Position	7%
	Entering a Parked Position	0.4%
	Turning Right	3%
	Turning Left	6%
	Making U-turn	14%
	Backing Up	8%
	Negotiating a Curve	1%
	Changing Lanes	1%
	Merging	0.4%
	Prior Corrective Action	-
	Other	14%
Driver Avoidance Maneuver	Object in Road	0.2%
	Poor Road Conditions	-
	Animal in Road	-
	Vehicle in Road	4%
	Non-Motorist in Road	-
	Hit and Run	8%
	No Driver Present	11%
	Other Avoidance Maneuver	0.1%
	Unknown	58%
	None	18%
	Phantom Vehicle	-
Corrective Action Attempted	No Driver Present	11%
	No Avoidance Maneuver	15%
	Braking	1%
	Releasing Brakes	-
	Steering	3%
	Braked and Steered	1%
	Accelerated	-
	Accelerated and Steered	-
	Other	0.2%
	Unknown	69%

Driver and vehicle statistics represent all light vehicles involved.

DOT HS 810 767
April 2007

U.S. Department
of Transportation

**National Highway
Traffic Safety
Administration**

www.ingramcontent.com/pod-product-compliance
Lightning Source LLC
Chambersburg PA
CBHW080258180526
45167CB00006B/2580